WARFARE IN PREHISTORIC BRITAIN

JULIAN HEATH

AMBERLEY

For my Family and Friends

First published 2009

Amberley Publishing Plc
Cirencester Road, Chalford,
Stroud, Gloucestershire, GL6 8PE

www.amberley-books.com

British Library Cataloguing in Publication Data.
A catalogue record for this book is available from the British Library.

ISBN 978 1 84868 369 3

Typesetting and Origination by Diagraf (www.diagraf.net)
Printed in Great Britain

CONTENTS

ACKNOWLEDGEMENTS

The first person that I have to thank is Peter Kemmis Betty, who originally accepted my idea for a book on prehistoric warfare in Britain and who gave me the chance to write about this fascinating but somewhat overlooked aspect of British prehistory. I must also express my gratitude to the many people who have provided me with photographs and information that have greatly helped in the presentation of this book. Many thanks therefore go to Richard Osgood, Rick Schulting, Don Benson, Marilyn Peddle, Mercedes Okumara, Ian Cartwright, Sigurd Towrie, Paul Backhouse, Anne-Rachael Harwood, Andrew Fitzpatrick, Mike Garner, Laura Pooley, Kathryn Charles-Wilson, Sally Jennings, Katie Anderson, Michael Craig, Palden Jenkins and Nikki Braunton. Further thanks go to Harry Fokkens, Rob Kruszynski, Chris Stringer, Barry Cunliffe, Mike Pitts, Joan Taylor, Phil Freeman and Mike Parker Pearson. Finally, I would like to say a big thank you to Margaret for proofreading the text and for providing some welcome suggestions as to how it could be improved.

INTRODUCTION

'Human war has been the most successful of all our cultural traditions'
(Robert Ardrey)

From the arrival of Caesar and the Roman Army in 55 BC to the death throes of the Jacobite rebellion at Culloden Moor on 16 April 1746, warfare has often cast its deadly shadow over British history. However, it is evident that warfare – taking 'warfare' to mean organised and lethal armed conflict between different groups – did not simply arrive in Britain with the formidable Roman legions. Indeed, as we will see through the course of this book, warfare in Britain appears to have a pedigree that at the least, reaches back some 6000 years into the Neolithic, and it is likely that warfare was present thousands of years before this in the Mesolithic and Late Upper Palaeolithic. Whatever the reality is as to the earliest occurrence of warfare in Britain, it can be justifiably argued that among the countless reminders left by the prehistoric people of Britain as mute but fascinating testimony to their lives, there are many which are indicative of this most terrible facet of humanity which gives lie to the word 'mankind'.

However, for much of the latter half of the twentieth century, warfare took something of a backseat in prehistoric studies (and to some extent, this still holds true). Therefore, 'peaceful' aspects of life such as trade, technology and religion came to the fore to provide the social framework on which reconstructions of prehistoric life were built, whilst the darker side of humanity, with which we in the twenty-first century are unfortunately all too familiar, was seen as having little impact on its development.

This somewhat biased interpretation of prehistoric life seems rather curious when it is considered that the evidence from archaeological sources strongly indicates that warfare was a significant feature of life in prehistoric Europe in general. As Helle Vankilde (2003: 127) has remarked in this regard, although trauma and organic weaponry are undoubtedly scarce, there is in fact, a considerable body of evidence pertaining to warfare: 'The number of prehistoric weapons, including fortifications, is huge, and iconographic representations of war and warriors in art and rituals supplement the picture. Skeletal traumata are, in fact, relatively frequent in European prehistory when it is taken into account that skeletons are often not well preserved, that skeletons are not routinely examined for marks of violence, and that much physical violence does not leave visible traces on the skeleton'.

Undoubtedly, it would be unfair to say that this evidence was ignored by archaeologists and indeed, warfare and warriors were not totally excluded from post-war models of prehistoric Europe. For example, the renowned mid twentieth-century Australian archaeologist, Vere Gordon Childe, described (1958: 144) the famous 'Beaker-Folk' of Copper Age Europe thus: 'they travelled fast in well-armed bands, their objectives were not only pastures and arable lands, but also raw materials for trade and industry'. Nevertheless, there is more than a little truth in the argument that warfare was often idealised by using terms such as 'unrest' and 'troubled times', rather than actual 'war', and by transforming warriors into heroic and somewhat aristocratic male figures, rather than violent killers who were responsible for death, destruction and terrible suffering (Vankilde 2003: 131-132).

The structural-Marxist approaches of the late 1970s and 1980s and post-modern archaeology of the late 1980s and early 1990s continued in a similar vein, with warrior elites sitting at the head of unequal societies, but with little mention of brutality, killing and warfare (*ibid.*). It could also be argued that the sites and weapons associated with warfare tended to be seen as interesting but static artefacts, which were analysed and typologically arranged, but were not really viewed as possible reminders of bloody conflicts in prehistoric Europe. Thus it could be said that not only was the possible impact of warfare on prehistoric society undervalued as a result of this approach, but once again, the terrible reality of warfare with all its attendant horrors was suppressed.

One notable post-War archaeologist who did present a more realistic view of the possible effects of warfare in prehistoric Europe was Marija Gimbutas. She argued that the Late Copper Age societies of eastern and central Europe had never recovered from a series of devastating invasions from the north Pontic Steppes (Chapman 1999: 103). Later, Gimbutas refined her chronology and theory and suggested that the Copper Age societies of the Balkans were originally peaceful and ruled by women, but these were subsequently transformed into violent, male-dominated societies by the successive waves of Kurgan invaders from the north Pontic Steppes (*ibid.*: 104).

Although Gimbutas' later theory has since been criticised and plausibly linked to the Soviet invasions that she witnessed as a child in Lithuania, she was one of very few post-war scholars who argued that warfare was of some significance among the prehistoric societies of Europe (*ibid.*). As Helle Vankilde (2003: 132) says, '[Gimbutas] treads in the footsteps of Childe with her emphasis upon warriors and migration, but her explanation contrasts with contemporaries in two ways: first, it has an inherent binary opposition of warlike maleness and peaceful femaleness ... Second [and most importantly], it contains an unusually direct reference to state of violence and warfare, which is considered fatal for human life and values of equality'.

In the late 1990's however, the situation changed as prehistoric warfare emerged to become a hugely popular topic in archaeology and increasingly, archaeologists began to look seriously at the effects that warfare may have had on the various societies of prehistoric Europe. As with all paradigm shifts in archaeology (and indeed, in other academic disciplines), it is likely that the nature of contemporary society had some bearing on this sudden and intense archaeological interest in prehistoric warfare. It may be that this interest reveals a social reaction to the many ethnic conflicts and genocides that erupted around the world in the late twentieth century (*ibid.*: 136). Furthermore, it is quite possible that the massive media coverage given to these wars forced archaeologists into a reappraisal of the evidence for prehistoric warfare and violence, and that unlike previously, archaeologists began to consider the true implications of this evidence (*ibid.*).

From within the archaeological community itself, an undoubted influence in changing attitudes towards prehistoric warfare was Lawrence Keeley's seminal book, *Warfare Before Civilization: the Myth of the Peaceful Savage* (1996). In this absorbing work, Keeley essentially argued that many anthropologists and archaeologists of the later twentieth century had 'artificially pacified the past' by ignoring or dismissing the significance of the plentiful evidence for primitive and prehistoric warfare that is present in both the ethnographic and archaeological records. Keely plausibly locates the origin of the 'neo-Rousseauian' (Jean-Jaques Rousseau was the influential eighteenth-century scholar who promoted the idea of the Peaceful and Noble Savage) views of these anthropologists and archaeologists in the immediate aftermath of World War II and feels (*ibid.*: 164) that the horrors and slaughter witnessed by both victors and vanquished alike 'encouraged a pervasive and profound odium for everything connected with warfare'.

Following a similar train of thought to Keeley, Roger Mercer (1999: 143) has said in regard to the role of warfare in European prehistory: 'European scholars, looking around themselves in the late decades of the twentieth century, can perhaps be readily forgiven, in the open vistas archaeology provides, for avoiding consideration of an aspect of human conduct that has brought so much protracted and intense pain to the entire continent for nearly a century of European civil war'.

Mike Parker Pearson (2005: 20) has also made the credible suggestion that the younger generation of post-processual archaeologists avoided the subject of warfare because they were reacting against their seniors' portrayal of the European Iron Age and earlier periods, which was based on the somewhat unreliable writings of the Classical authors.

While there can be no denying the importance of *War Before Civilization*, not all have agreed with Keeley's warlike vision of the past and his book has received mixed reviews from scholars. Perhaps the fiercest critic of Keeley (and vice versa) has been the noted social anthropologist, Brian Ferguson, and there have been several heated exchanges between the two scholars in regard to the question of warfare in primitive and prehistoric societies. In response to the view of prehistoric warfare taken by Keeley in *War Before Civilization*, Ferguson (2000: 159) has said: 'The question of the antiquity of war has been raised but clouded by Keeley (1996), whose rhetoric exceeds his evidence in implying war is as old as humanity'. In a similar critical vein to Ferguson, John Chapman (1999: 102) has suggested that Keeley's argument that prehistoric warfare was just as significant as that of state led warfare is undermined by the fact that he presents only a dozen examples of warfare from 30,000 years of prehistory. Another critic is Mike Parker Pearson (2005: 25), who accuses Keeley of twisting and mangling the anthropological and archaeological evidence to support his theory.

Nonetheless, although Keeley's book is rather polemical in its approach and is not without its flaws, its main message that all war is total and unlimited – whether it be 'tribal', 'civilised', 'primitive' or 'prehistoric' – is perhaps its greatest strength (Carman and Harding 1999: 5). Ultimately, Keeley forced both anthropologists and archaeologists to reconsider the role of warfare in primitive and prehistoric societies and challenged them to continue to believe in the doctrines of the pacified past, despite the fact that the arguments and evidence indicated otherwise (*ibid*.).

This book then, is just one of many studies of ancient (and more recent) warfare that have been published in the last decade or so, and it can be seen as part of the current trend being followed by both North American and European archaeologists who are justifiably attempting to un-pacify the past (Armit *et. al.* 2006: 1). Indeed, there can be little doubt that prehistoric warfare is now an important and valid area of archaeological enquiry which is increasingly showing that as much as we may like to think otherwise, the prehistoric world was not one in which communities always lived peacefully side by side, in harmony with nature and each other.

CHAPTER 1

THE LATE UPPER PALAEOLITHIC AND MESOLITHIC C.10,000-4000 BC

This book begins in Cheddar Gorge on the southern edge of the Mendip Hills in Somerset, for it is here at the famous site of Gough's Cave that we perhaps have the earliest evidence for the occurrence of warfare in prehistoric Britain. Although this evidence does not provide concrete proof of prehistoric warfare, it is certainly intriguing in this regard.

GOUGH'S CAVE IN THE LATE UPPER PALAEOLITHIC: A SITE OF WARFARE CANNIBALISM?

With its spectacular interior and stunning location amongst the towering limestone cliffs of Cheddar Gorge (*1*), it is unsurprising that Gough's Cave has acted as a tourist magnet since the early twentieth century. However, it is evident that people were visiting the cave many thousands of years before this, as various archaeological discoveries (many of them made when the entrance of the cave was being cleared to allow access into its interior) have provided firm evidence of prehistoric activity at this site. Some of this activity relates to the use of the cave during the Late Upper Palaeolithic some 12,000 years ago, when semi-nomadic hunter-gatherer communities occupied the landscapes of Britain, living off the wild resources that nature provided for them.

Amongst the archaeological material found at the cave is a collection of human bones that provide us with strong evidence of an episode of cannibalism among these Late Upper Palaeolithic hunter-gatherers. As we will also see below, there are also certain aspects of this evidence which indicate that this was probably an act of 'aggressive exocannibalism', which involved the eating of dead enemies or outsiders (Fernandez-Jalvo 2003: 593). As Larry Barham (1999: 78) points out, this type of cannibalism was sometimes carried out in order to terrify enemies and to create a fearsome reputation. Those who practiced warfare cannibalism also often did

1 Cheddar Gorge (Palden Jenkins)

so in order to humiliate the enemy (Keeley 1996: 106) and because they believed that they would absorb the vital energies and personal attribute of their enemies if they consumed their flesh (Thorpe 2003: 158). Although some may dispute that our prehistoric predecessors in Britain were incapable of committing such acts, not only do several further archaeological discoveries from the prehistoric period point in this direction, anthropologists have also documented the occurrence of warfare cannibalism in several non-state societies around the world.

CANNIBALISM IN THE ARCHAEOLOGICAL RECORD

Gran Dolina

The earliest possible case of European cannibalism yet discovered comes from the Gran Dolina cave in the Sierra de Atapuerca, Spain, where the remains of at least six archaic humans probably of the type known as *Homo heidelbergensis* (though the remains have also been designated as a new and earlier species – *Homo anteccesor*) were found at a site dating to some 780,000 years ago (Fernández-Jalvo *et. al.* 1999; Lukaschek 2000/2001: 14). Cannibalism is suspected at the site as the bones of the people were found mixed together with animal bones and stone tools and both sets of bones display similar cut-marks and breakage patterns that relate to the removal of flesh and marrow. This evidence has led to the conclusion (Fernández-Jalvo *et. al.* 1999: 620) that the individuals buried in the Gran Dolina Cave were the victims of another group, who brought their bodies to the site to be butchered and eaten alongside various animals. However, although it seems to suggest otherwise, it may be that the evidence at Gran Dolina represents the occasional consumption of people by members of their own group after death, rather than an act of warfare cannibalism (Thorpe 2005: 7).

At Fontbrégoua cave in south-eastern France, we have more equivocal evidence for aggressive exocannibalism (Lukaschek 2000/2001). Here, three separate clusters of human bones and ten clusters of animal bones were discovered and these probably relate to the use of the cave as a temporary camp during the fifth and fourth millennia BC. Again, there were cut-marks and breakage patterns on both human and animal bones which implied that both human and animal carcasses had almost certainly been processed for food using the same butchering techniques. In addition to this evidence, there were signs that the Neolithic people buried in the cave were indeed the victims of an enemy group. In Cluster H3, there were the remains of at least six individuals (with 3 adults and 2 children identified) who had been killed in a single event and who appear to have been butchered with a broken stone axe that was found with them (*ibid.*: 22). Human skulls found in the cave also seem to have been casually discarded in a rubbish pit (*ibid.*: 24). The human skeletal material found at Fontbrégoua could of course represent some type of non-aggressive ritual act, although this seems unlikely. As Villa (in Billman *et. al.* 2000: 165) has argued, 'If the human bones at Fontbrégoua indicate secondary burial, we may conclude that the Fontbrégoua people hunted, herded, and butchered but did not eat food animals and that they gave secondary burial to boars, sheep, roe deer, badgers and marten'. A similar argument could be applied to the evidence from the other sites mentioned previously.

Convincing evidence for warfare cannibalism has been found in south-west America at several sites of the prehistoric Anasazi people (e.g. Canyon Butte 3, Mancos Canyon and LA 4528, to name a few). One of the best known sites is 5MT10010, which is located on the Cowboy Wash flood plain in the Mesa Verde region of south-western Colorado (Billman *et. al.* 2000). Site 5MT10010 consists of three pithouses that represent one of ten habitation sites of a 'Pueblo III' community dating to 1125-1150 AD. Two of the three pithouses provided compelling evidence of cannibalism. This took the form of deliberately broken and cut-marked bones, stone tools with human blood on them and most tellingly, a human coprolite that contained traces of myoglobin, which is only found in the cells of human skeletal and cardiac muscles.

ETHNOGRAPHIC ACCOUNTS OF CANNIBALISM

Having briefly considered the archaeological evidence for possible warfare cannibalism in the prehistoric period, we will turn now to the anthropological record which contains many examples of this practice. While some of these should rightly be treated with caution given that they come from unreliable sources and were undoubtedly used as a form of propaganda against 'savages' as a means of justifying the exploitation of foreign lands by Europeans, not all can be discarded as such.

Some of the most famous accounts of cannibalism come from the Spanish conquistadors of the sixteenth century. They recorded that some of the war captives who were ritually murdered in their thousands during the religious ceremonies carried out by the famous Aztecs were also the victims of cannibalism. There can be little doubt that a number of these accounts cannot be relied upon and provide us with a good example of the propaganda mentioned above. However, as Timothy Taylor (2001: 9) notes, large numbers of weathered and broken bones have been discovered outside Aztec temples, from where they must have been swept, and so, it seems that not all the conquistadors' accounts were spurious.

In southern Amazonia the warriors of the Wari' people removed the heads, arms and legs of the enemies (who like animals, were viewed as prey) they had killed in battle and transported them in baskets to their villages, where they were roasted and eaten (Viliça 2000: 90).

From Lesotho in South Africa, there are ethnographic accounts of a potion called *Diretlo* or *Ditlo* that was believed to boost courage and provide protection from harm and which could only be made from the flesh of a war captive or outsider, while in other parts of Africa, enemy hearts were pounded into a power-giving potion (Davis in Lukaschek 2000/2001: 7). Similarly, the famous Zulu drank a soup that was made from various 'powerful' parts of a victim (e.g. penis, right forearm, breastbone), in order to provide strength in battle (Keeley 1996: 106).

There are also several accounts that the Zande people of Central Africa were cannibals and although many of these accounts are dubious, it is highly likely that enemies killed in battle and criminals were eaten, though probably not in great numbers (Evans-Pritchard 1960: 251-252).

Many of the tribes and chiefdoms of Colombia are also reputed to have eaten large numbers of the enemy dead and war captives, with one account recording the fattening up of captives in special enclosures before they were eaten, and another telling of an occasion when a chief and his followers consumed flesh from 100 enemies in one day (Keeley 1996: 103).

THE EVIDENCE FROM GOUGH'S CAVE

The first hints of prehistoric cannibalism at Gough's Cave were found by R.F. Parry during the excavations undertaken at the cave in 1928 and 1929. The fragmentary remains of two adults and a child were discovered and these included a large lower jaw from an adult aged about 30 (Parry *et. al.* 1930: 57) Intriguingly, the lower jaw displayed cut marks from a flint knife on its inner and outer surfaces and these may relate to the removal of the tongue for consumption (Barham 1999: 77).

Of course, a single jaw displaying cut marks provides rather shaky support for the idea that people were eaten inside Gough's Cave by their fellow humans at some point during the Late Upper Palaeolithic. However, during 1986-1987, limited excavations were undertaken on a small area threatened by flooding and erosion (2) and subsequently, a mixture of animal bones, stone tools and several human bones displaying further cut marks were discovered (Currant *et. al.* 1989). A detailed analysis (Andrews and Fernández-Jalvo 2003) of the skeletal material found at Gough's Cave during the 1986-1987 season, and later excavations at the site up to 1992 revealed that both human and animal bones display numerous cut-marks, percussion marks and 'peeling' (the deliberate snapping of bones, probably to extract marrow). Two adults, two teenagers and one child aged about three years are represented by the human skeletal material from Gough's Cave, while Red deer, wild cattle and horses are found among the animal bones, with red deer and cattle predominant (Lukashek 2000/2001: 21).

Included among the human skeletal material is an adult 'calvaria' or skull cap with cut marks on either side and around the top (*colour plate 1*), and these show that not only had the jaw been removed, but also that the scalp had been cut away (Barham 1999: 80). In addition to these cut-marks there is an indication that this individual had suffered a serious blow above the right eye

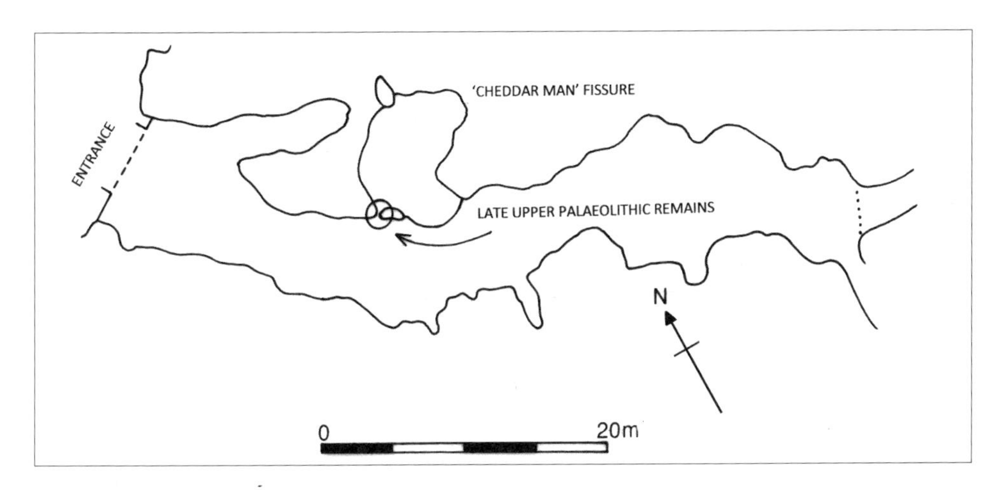

2 Plan of Gough's Cave (redrawn after Currant *et. al.*)

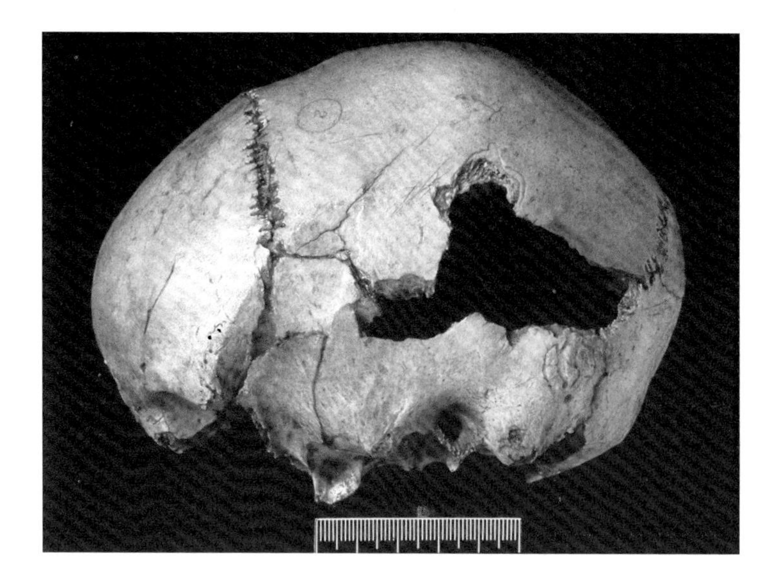

3 Child's skull with cutmarks from Gough's Cave (Natural History Museum)

and while this may simply be the result of an accident, it is possible that it was an injury received in an episode of warfare (*ibid.*). It is worth noting that when discovered, this skull cap was lying upside down and was filled with red deer bones and appears to have been deliberately trimmed to make a 'bowl' (Currant in Barham 1999: 80). The skull caps of the adolescent and child also display cutmarks (3) (Andrews and Fernández-Jalvo 2003: 74), as do neck vertebrae found at the site, which points to the decapitation of an individual as he or she was lying face down on the floor of the cave (Korn in Lukaschek 2000/2001: 21). It is also particularly noticeable that both the human and animal jaws at Gough's Cave are heavily broken and cut in a way which indicates the deliberate removal of tongues and it is evident that human and animal ribs display cut and chop marks which relate to deliberate entry into the thoracic (chest) cavity (Andrews and Fernández-Jalvo 2003: 79, 76).

This skeletal evidence then is very suggestive of cannibalism taking place at Gough's Cave at some point in the final stages of the Upper Palaeolithic. Whether this reveals that the occupants of the cave at this time were attacked, killed, butchered and eaten by a rival group of hunter-gatherers (Barham 1999: 79) is of course, open to question. It could be that the Gough's Cave evidence points to an act of 'reverential funerary endocannibalism', which is not unknown among small-scale societies, and involved eating members of one's own family or group (Taylor 2001: 10), or even perhaps survival cannibalism stimulated by a severe shortage of food. However, the casual discarding of the human bones and their mixing with the animal ones seems to mitigate against survival cannibalism, and indicates a lack of respect for the dead. Furthermore, the indication of decapitation and the individual who not only appears to have been scalped but also probably received a serious blow to the head, suggests that it is more likely that the occupants of Gough's Cave were indeed eaten by their enemies in an act of warfare cannibalism.

OTHER POSSIBLE EVIDENCE FOR LATE UPPER PALAEOLITHIC AND MESOLITHIC WARFARE IN BRITAIN

Not far from Gough's Cave in the Mendips is the cave known as Aveline's Hole, which functioned as an Early Mesolithic cemetery some 9000 years ago. Around 100 skeletons were discovered in the cave during various excavations, making this the largest collection of Mesolithic human remains found in the British Isles and also one of the largest from the continent (Schulting and Wysocki 2002: 255). Intriguingly, cut-marks were reported in earlier reports on the skeletons. Although these cut-marks cannot now be located either on the surviving skeletal material (much of which was destroyed in an air raid on Bristol in 1940) or on photographs, as Rick Schulting and Michael Wysocki (*ibid.*: 262) rightly remark, they are worth bearing in mind in light of the evidence from Gough's Cave. It should also be noted that on a photograph of a skull from Aveline's Hole, a perforation can be seen (Schulting in Thorpe 2006: 143) and it is possible that this hole is related to a violent incident or an injury received in an armed clash between hunter-gatherers.

Similar evidence has been found at Caldey Island off the south-east coast of Wales, where a probable male skull dating to *c.*8500 BC was discovered in a cave known as Ogof-yr-Ychen ('cave of the oxen') and this displayed a healed depressed fracture (*ibid.*).

Near the village of Whaley in Derbyshire further ambiguous evidence for armed conflict between prehistoric hunter-gathers was found in a rock-shelter. Here, an adult skull cap was found in an Upper Palaeolithic occupation level that included bone, antler and flint artefacts (Brothwell 1961). One of the excavation team believed that the condition of the skull indicated cannibalism and violent death and he says (Armstrong in Brothwell 1961: 115): 'The maxillae [face] and base of the skull have apparently been deliberately removed, presumably for access to the brain [and] Definite evidence of death by violence is provided by the square holes which pierce the top and sides of the skull, believed to have been inflicted by spears of wood or bone'. However, it may be that Armstrong jumped too readily to macabre conclusions. As Brothwell (*ibid.*) has said, 'After carefully examining all the holes in the vault [of the skull], I can see no reason why the apertures should not be explained by the considerable fragmentation which the vault has undergone'.

Finally, we return to Gough's Cave and the skeleton that was discovered in 1903 by workmen digging a drainage ditch (2) when the site was being converted into a tourist attraction by R.C. Gough, the owner of the cave (Tratman 1975: 8). The skeleton belonged to a male who had lived in the Early Mesolithic around 8000 BC and who was rather short of stature at around 5ft tall (Thorpe 2006: 143; Wilson 2001: 53). 'Cheddar Man', as he was subsequently named, has since gone on to become rather famous, not least because it was found that somewhat ironically, Adrian Targett, a history teacher in a local school is directly descended from him.

It seems that Cheddar Man had not ended his days in peace, as an examination of his skull showed that he had received several serious blows to the head, one of which landed between his eyes and caused a tiny piece of bone to become dislodged from his skull. As a result, a large abscess formed (*colour plate* 2), which must have given him a permanent and painful headache and it is very likely that this subsequently became seriously infected and eventually killed him (*ibid*.: 146).

Of course, the meagre and somewhat ambiguous evidence described above does not prove that warfare existed amongst Britain's Late Palaeolithic and Mesolithic hunter-gathers as it may relate to unknown religious practices, violent encounters between individuals, or even accidents or post-depositional damage. However, aside from the fact that anthropologists have recorded warfare amongst many hunter-gatherer societies, further discoveries from prehistoric hunter-gatherer sites around the world caution us against refuting its existence in Late Upper Palaeolithic and Mesolithic Britain.

ARCHAEOLOGICAL EVIDENCE FOR PREHISTORIC HUNTER-GATHERER WARFARE

Some of the most graphic evidence for prehistoric hunter-gatherer warfare comes from the Late Palaeolithic cemetery known as site 117 (dating to *c*.13,000 BC) which is located on the east bank of the Nile near Gebel Sahaba in the Sudan (Wendorf and Schild in Thorpe 2005). The cemetery contained 59 burials of the 'Qadan culture', with people buried in simple oval pits covered by flat slabs of stone (Hoffman 1984: 91), and a dramatic and violent event is suggested by the evidence found within the graves. Twenty-four out of the 59 skeletons had chert projectile points (with 110 points found) embedded in the bones or in the grave fill and some were even found inside skulls (Thorpe 2005: 8; Hendrickx and Vermeersch 2000: 30; Hoffman 1984: 97). It is probable then, that Cemetery 117 bears witness to a Late Palaeolithic massacre in the Nile Valley and the cut marks seen on some of the bones, the existence of multiple burials (with 8 people buried in one grave) and the fact that some 50 per cent of the people buried here were women and children lends considerable support to this idea (Hendrickx and Vermeersch 2000: 30).

At the Final Palaeolithic site of Wadi Kubbaniya in southern Egypt, two flint bladelets were found between the ribs and backbone of an individual and were probably attached to a projectile that had killed this person (Wendorf & Schild in Thorpe 2005: 8).

Further evidence for armed conflict in the Late Palaeolithic comes from the San Teodoro cave in Sicily and the Grotta dei Fanciulli ('cave of the infants') on the Italian mainland (Bachechi *et. al.* 1997). At the former, the tip of a flint projectile (probably an arrowhead in the opinion of the excavator) was found embedded in the pelvis of a probable female skeleton, while at the latter, another probable arrowhead was found in one of the thoracic vertebrae of one of two children who had been buried together.

At Franchthi Cave in the Gulf of Argos in Greece, which was occupied from the Upper Palaeolithic to the Neolithic, another possible victim of hunter-gatherer warfare was discovered. During the second season of excavations at the site, a young man (25-29 years) from the Early Mesolithic was found in a shallow pit and it was evident that he had died from a number of violent blows to his forehead (Cullen 1995: 275).

Strong evidence for prehistoric hunter-gatherer warfare has also been found in southern Scandinavia in the burials of individuals belonging to the Late Mesolithic Ertebølle culture whose well-preserved sites have provided us with remarkable snapshots of hunter-gatherer life in northern Europe.

At the site of Vedbæk on the island of Zealand, some 20km north of Copenhagen, an Ertebølle cemetery dating to around 4100 BC was discovered and Grave 19 contained the well preserved skeletons of two adults (one aged 25-30 and the other 35-40 years old) and a one year old infant (Albrethsen and Brinch Petersen: 1976). Although the sex of the adults could not be conclusively determined, it is probable that they represent a male and female, and therefore we are probably looking at a family burial. It is clear that the youngest of the adults (the 'male') had died violently, as there was a bone point wedged between his neck vertebrae which would have caused instant death. But how had the woman and child died? Interestingly, a flint knife lay immediately below her jaw and thus Thorpe (2000: 13) has suggested that the woman and child may have been sacrificed to accompany the man into the afterlife, after he had been shot in an armed conflict with a neighbouring group. Possible evidence of warfare cannibalism was also found at the site, with long bones that appear to have been broken in order to remove the marrow found in a refuse layer and Grave 11, also yielded a human skull cap (Albrethsen and Brinch Petersen 1976: 5, 22).

At Bäckaskog in Sweden, the grave of an adult female was discovered and she had probably been killed by a bone point that was fired into her chest (*ibid.* 24), whilst a middle aged male found at Stora Bjers in Gotland had a bone point embedded in his pelvis (Newell *et. al.* 1979: 39-40).

Further evidence for armed conflict among Ertebølle communities has come from Skateholm in southern Sweden, with projectile points found in the ribs of one individual and in the pelvis of another (Price 1985: 352). There were also indications of cannibalism at the site with finds of scattered and burnt human bones and a deliberately split long bone (*ibid.*). At the site of Dyrholmen in Jutland, the bodies of at least nine people were found and cut-marks could be seen on skulls, suggesting scalping, and long bones were also cut-marked and fractured, again implying that people had removed the marrow (Thorpe 2000: 13).

Bones that had been deliberately broken were also found at the coastal site of Møllegabet in Denmark and a male seems to have had his jaw broken and his teeth removed, perhaps by his killer, who may have incorporated the teeth into a bodily ornament (*ibid.*). The body of a young man placed in a dugout canoe was found at Møllegabet and he had a healed wound on his head that had been caused by a severe blow (Whittle 1996: 198).

In addition to the above evidence, Pia Bennicke's study of prehistoric cranial injuries in Denmark revealed that the greatest proportion of injuries came from the Mesolithic period, suggesting that clubs were a favoured weapon during this time (Thorpe 2000: 12). Examples include the male burial from the Ertebølle site at Korsør Nor harbour who had a deep but healed wound caused by a club and healed wounds were also found on the skulls of an elderly male and an adult of unidentified sex (Thorpe 2003: 155). A female buried with a young child at Vedbæk also had a healed fracture of the skull, apparently caused by a blunt instrument (*ibid.*: 156).

In Brittany at the important Late Mesolithic site of Téveic in the Bay of Quiberon, a young adult male found in one of the ten graves at the site (which contained 23 individuals) had the fragments of two microliths embedded in his sixth and eleventh thoracic vertebrae and also displayed a healed fracture of his jaw (Schulting 1996: 339-340). Another young adult of indeterminate sex, also showed signs of facial injuries and had a hole in his skull that had only partially healed (Persson and Persson in Thorpe 2005).

Evidence for armed conflict between European hunter-gatherers has also been found in the large cemetery complexes found in the Dnieper Rapids region of the Ukraine which date to the transitional period between the latest Palaeolithic and first Mesolithic, *c.*10,400-9200 BC (Lillie 2004). At the cemetery of Voloshkoe (Final Palaeolithic), there was an individual with a bladelet embedded in his/her atlas bone, while a male, who also had his hands cut off (though whether by an enemy is not known) before being buried, had a bladelet lodged in his breastbone.

Although the evidence is not conclusive, two other individuals in the cemetery had probably been killed by projectile points and like the other victims of violence found here, they might have been killed by an attacking force of enemy archers.

The cemetery of Vasilyevka III (probably Mesolithic in date) contained individuals of both sexes who had been shot by 'microliths' (small pieces of worked flint or chert) that had functioned as arrowheads. One was found between the ribs of a female aged 18-22, while a male aged 25-35 had a microlith embedded in his lumbar vertebrae and he had obviously been shot from behind. An individual of unknown sex also had a microlith fragment and two spear-point fragments embedded in his/her skeleton, suggesting attack by a number of individuals, while a female aged around 20 had a slotted bone point in her spine.

Malcolm Lillie (*ibid.*: 90, 95) feels that it is possible that the skeletal evidence from the Ukraine reveals that the hunter gatherer communities of the Dneiper Rapids were involved in armed conflicts which arose because of competition over the prime fishing territories in this region. Following a similar train of thought, Nick Thorpe (2005; 11) has suggested that Mesolithic groups in coastal areas (such as those of Brittany or Scandinavia) may have fought over territories where shellbeds were an important resource, with some groups perhaps attempting to mark these territories by raising shell middens in the landscape. Perhaps then, Britain's Late Upper Palaeolithic and Mesolithic hunter-gatherer communities also clashed over resource-rich areas where there were prime concentrations of fish, birds and other wildlife and that places such as river mouths may have been 'flashpoints' in this regard (Gat 2006: 64).

It could be argued that such resource-driven warfare could never have broken out in Late Palaeolithic and Mesolithic Britain because the small size of populations during these periods would mean that there would be 'plenty to go around'. However, evidence from the anthropological record indicates that we should be wary of assuming that this was the case. For example, in North America, Plains Indians tribes are known to have fought over herds of bison (buffalo) in early historic times, at the end of the sixteenth century AD (Newcomb 1950: 324).

In Australia, aboriginal tribes often fought over both water and food resources (Gat 2006: 63) though of course, it is highly unlikely that Britain's last hunter-gatherers clashed over water resources. Azar Gat (*ibid.*: 64) has also argued that rising population levels among hunter-gatherers could lead to depletion of wild game and thus competition and conflict over food would become an inseparable part of their life, particularly in areas where there were dense and unevenly distributed concentrations of food.

Of course, to argue that hunter-gatherer warfare in Late Palaeolithic and Mesolithic Britain was solely over resources is too simplistic, and as with subsequent prehistoric periods, there must have been other reasons as to why people sought to kill each other in armed conflicts. Some suggestions as to what these might have been will be briefly discussed in the conclusion of this book.

One of the most famous (or infamous) sites which has strong evidence pointing to hunter-gatherer warfare is Ofnet Cave in Bavaria. Here, the skulls and vertebrae of 34-38 individuals of the Late Mesolithic (*c.*6500 BC) were found in 'nests' or clusters in two pits close to the cave entrance during excavations in the early twentieth century (Thorpe 2003; Orschiedt 2005). Many of the skulls belonged to children (with some infant skulls also), while two thirds were female. Red ochre had been poured over the skulls and some 200 pierced red deer teeth and over 4000 shells were placed with the skulls. Over half of the skulls displayed serious injuries, although it was the smaller number of male skulls that displayed the most wounds and it is likely that the injuries were caused by stone axes or clubs of some sort. Cut marks (which may be related to scalping) were also present on some skulls and also on many of the vertebrae, revealing decapitation.

Taken together, the evidence from Ofnet is curious as it is clear that some care was taken in the burial of the skulls, but it is plainly evident that they belong to individuals who had died very violently. So who had been responsible for their burial? Thorpe (2000: 9) has plausibly suggested that they represent the trophies from a head-hunting expedition which were given a ceremonial burial in the cave by those who had taken them, rather than one carried out by the remaining men of the community (who may have been away hunting when their loved ones and friends were brutally killed), as it does seem unlikely that they would choose to bury the heads and not the bodies of the dead.

As Phillip Walker (2001: 573) has pointed out, the evidence found at Ofnet 'is important because it shows that the development of sedentary agricultural communities is not a prerequisite for organized, large-scale, homicidal activity'. In other words, warfare was a feature of life long before prehistoric people decided to settle down in villages and begin farming.

MESOLITHIC AND LATE UPPER PALAEOLITHIC WEAPONRY IN BRITAIN

Archery Equipment

We are unsure as to when the bow first appeared in Europe, but the skeletal evidence strongly suggests that it was at least as early as the Late Upper Palaeolithic and it is not impossible that the bow was invented several thousands of years before this. There may even be depictions of humans killed by arrows

4 Possible depiction of a human killed by arrows from Upper Palaeolithic cave on the Continent (redrawn after Bachechi *et. al.*)

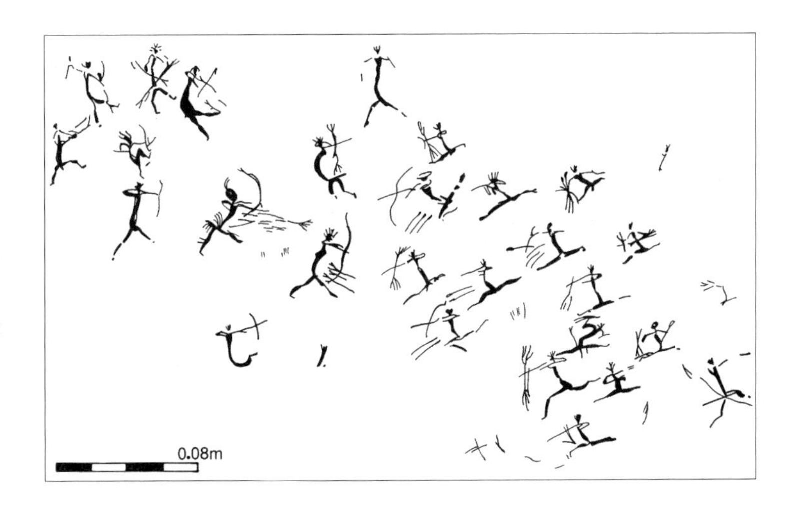

5 Mesolithic/ Neolithic rock-art from the Spanish Levant depicting archery battles (redrawn after Nash)

6 Rock art image of archer holding wounded or dead comrade in his arms – Spanish Levant (redrawn after Nash)

from some of the superbly decorated and highly intriguing Upper Palaeolithic cave-sites on the continent (4). However, it must be stressed that this evidence is far from conclusive, as the 'arrows' may actually be spears. It has also been rightly argued (Thorpe 2003: 152) that we cannot determine whether these depictions represent wished-for killings, magical killings or the actual deaths of people who had been killed by archers. Also, it is quite possible that the figures depicted are not human, and rather, they may be creatures that belong to the supernatural world of the Upper Palaeolithic communities who left them on the walls of their caves for us to puzzle over.

We should also consider here the fascinating rock-art found in the Gasulla and Valltorta gorges in the Spanish Levant (Nash 2005), which depicts opposing groups of prehistoric hunter-gathers engaged in archery battles (5). Depictions of what appear to be executions, gladiatorial scenes and a image of an archer holding a dead comrade in his arms – which is remarkably poignant for all its simplicity – can also be seen (6). Although recent opinion now largely favours a Neolithic date for these figures, it has been argued (*ibid.*: 79-80) that even if the figures do date to the Neolithic they still probably represent people following a hunter-gatherer lifestyle, as hunting and food gathering are also depicted alongside the warring archers.

Similar rock-art scenes depicting battles between prehistoric hunter-gatherers can be seen in other parts of the world, and examples include those painted by Aboriginal tribes in Arnhem Land, Australia (Taçon and Chippindale 1994), the San Bushmen of southern Africa (Campbell 1986), and by Plains Indians in southern Alberta, Canada (7) (Keyser 1979).

At present, we have no human skeletal evidence which reveals that the bow was used in warfare by Britain's Late Upper Palaeolithic and Mesolithic hunter-

7 Prehistoric rock art from Alberta, depicting battles of the Plains Indians (redrawn after Keyser)

gatherer communities. However, it should be remembered that human remains from these periods are few and far between in Britain and thus it is perhaps not surprising that we have struggled to find evidence for the use of the bow as a weapon of war. In addition to this, the skeletal and iconographic evidence cursorily examined above suggests that it would be somewhat strange if armed conflicts involving archers did not take place, at least on occasion, among Britain's Late Upper Palaeolithic and Mesolithic hunter-gatherers.

Unfortunately, we have not yet found any prehistoric bows earlier than the Neolithic in Britain. However, though there have been rare discoveries of both bows and arrows from Mesolithic contexts on the continent, which provide us with a fascinating glimpse of the weapons that the prehistoric hunter-gathers of Europe favoured not only in hunting but also in warfare.

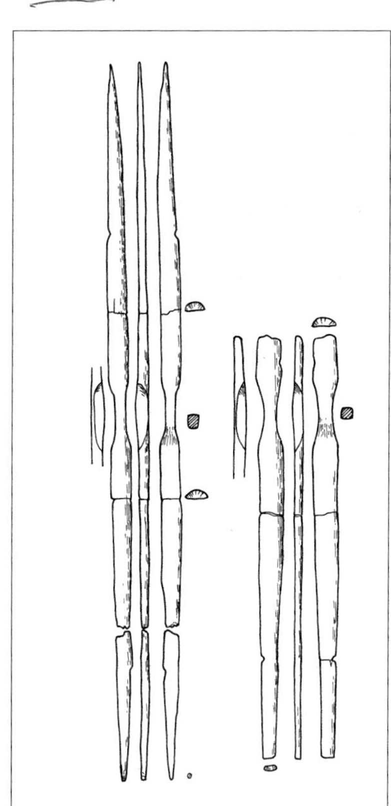

Right: 8 Mesolithic bows from Holmegaard IV (redrawn after Clark)

Below: 9 Mesolithic arrow from Loshult, Sweden (redrawn after Smith)

Two finely preserved bows were found at site of Holmegaard IV on the island of Zealand in Denmark (Clark 1963: 63) and although one was missing its ends, they were of a similar length (154 and 180cm) and design, with plano-convex limbs (*8*). Similar longbows have been discovered at the sites of Ageröd V and Ringkloster (Mithen 1997: 97).

Turning to arrows, there is the remarkable find of around 100 arrowshafts from the reindeer hunter's site at Stellmoor (*c*.8500 BC) in Germany, with two arrow shafts still retaining parts of their original flint arrowheads (Clark 1963: 62). In addition to arrows and the huge number of reindeer bones found at the site (some 18,000), the ends from two wooden bows were also found (Barton 2005: 135).

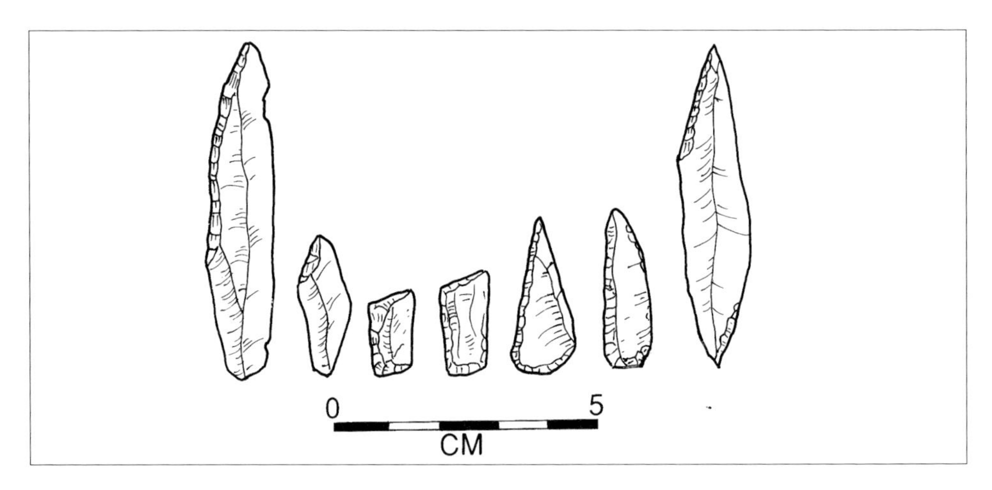

10 Mesolithic microliths (redrawn after Edmonds)

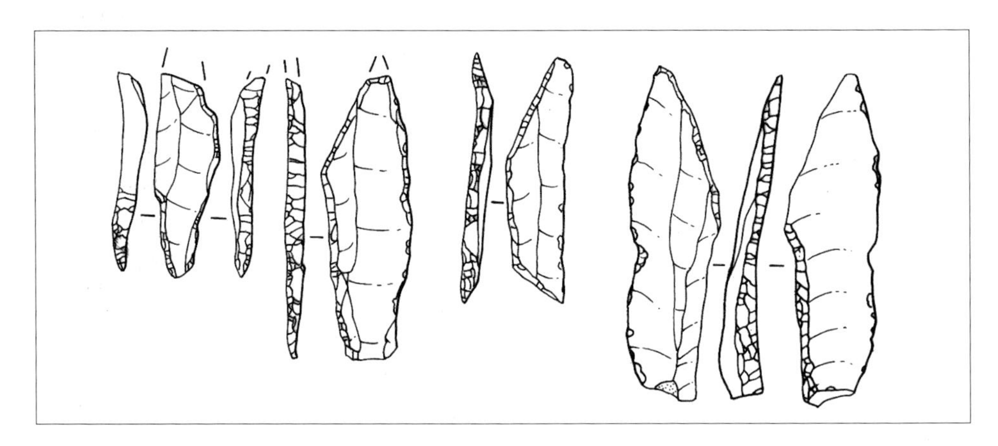

11 Late Upper Palaeolithic flint points (redrawn after Smith)

From Loshult in Sweden, there is an almost complete arrow shaft of pine wood, which still had its flint tip and barb in place, and the resin or pitch (probably obtained from birch bark) which had helped secure the microliths into the shaft (*9*) (Waddell 2000: 16).

At the famous British Mesolithic site of Starr Carr in the Vale of Pickering a microlith was discovered with part of its resin mounting still attached to it (Clark 1954: 102). Microliths are one of the defining characteristics of the Mesolithic (*10*), and while it is very likely that they were used for other purposes such as processing plant materials or as drilling implements (Mithen 1997: 97) it is undoubted that they were used mainly as arrowheads for hunting, as there have been many finds of microliths still embedded in animal bones. However, as we have seen with the skeletal evidence described briefly above, it is evident that they were also used to kill humans as well as animals. In regard to arrowheads from the British Late Upper Palaeolithic, various flint points that are not unlike the microliths of the Mesolithic in size and form probably served as armatures for projectile weapons (*11*).

Other weaponry

The skeletal evidence for trauma from Mesolithic societies clearly shows that although they may not have been designed primarily as weapons of war, clubs and axes were used in combat and the evidence found at Ofnet Cave provides a graphic example of the brutal damage that could be caused by such implements. As we have seen at Gough's Cave, Cheddar Man seems to have died as a result

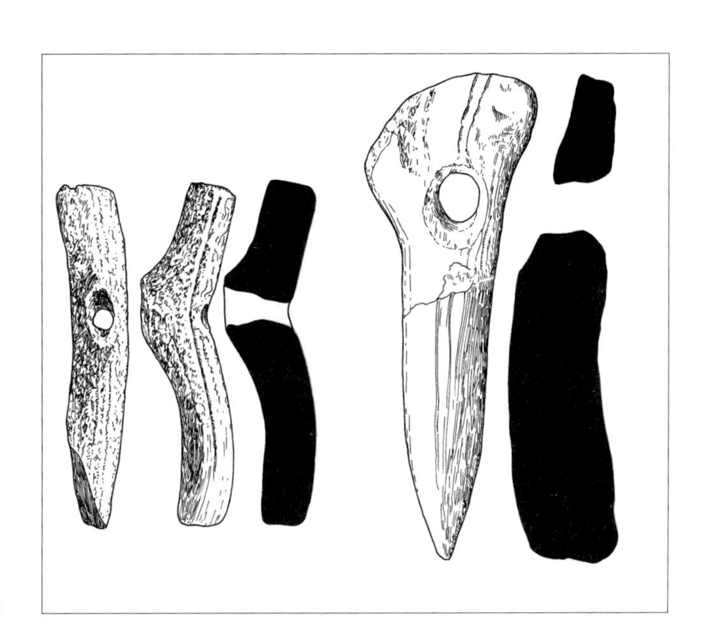

12 Antler mattocks from Meiklewood and Starr Carr (redrawn after Clark)

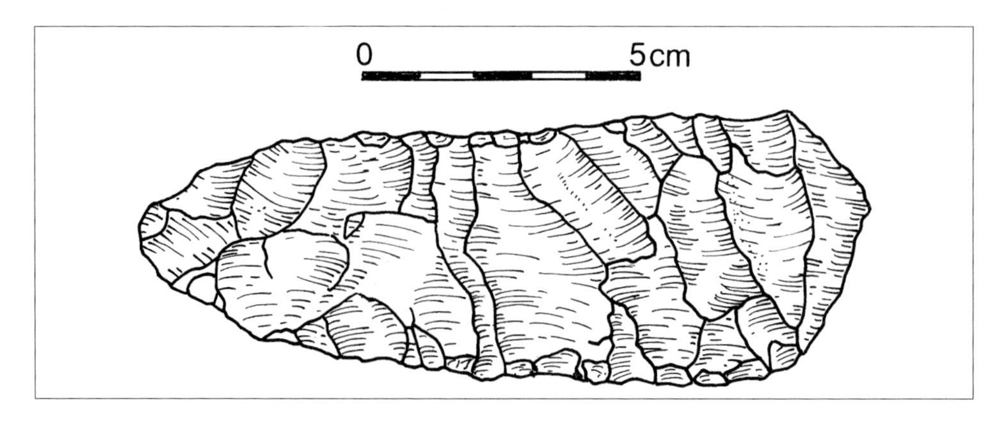

13 Mesolithic flint axe (redrawn after Edmonds)

of a blow to the head that was probably administered by a club of some sort and one of the adults from the Late Upper Palaeolithic level also appears to have been hit in the head with a club. Antler was a favoured material used for making 'mattocks' or clubs in the Mesolithic and examples have been found at various sites in Britain such as Starr Carr and at sites of the 'Obanian Culture' in Scotland such as Risga and Meiklewood (*12*). It is likely that simple wooden clubs were also used by both Mesolithic and Late Upper Palaeolithic societies in Britain.

The ethnographic record provides us with some support for the idea that organic weapons were used in warfare by Britain's late hunter-gatherer societies. A few examples among many are provided by the *Patu Potu* war clubs of the Maori, made from wood or whalebone, the wooden war clubs of the Fulani people of Nigeria and the *iwisa* or 'knobkerries' of the famous Zulu (Osgood 2000: 25).

Flint axes, which are often seen as one of the defining features of the Neolithic, were also used by the Late Mesolithic societies of Britain (*13*), along with perforated sandstone implements (*14*) which may perhaps be maceheads (Edmonds 1995: 26). It is interesting to note that disc-shaped maceheads that are similar in appearance to the perforated sandstone implements from Mesolithic Britain, were made by people of the Predynastic Naqada culture in Egypt and these were symbols of power (Midant-Reynes 2000: 50) that were also used in warfare.

Spears tipped by stone, bone and antler would also have been used by both Mesolithic and Late Upper Palaeolithic societies in Britain and although spears were undoubtedly vital hunting implements, they could certainly have doubled as efficient weapons in armed conflicts, if they were so needed. It is likely that simple wooden spears without flint or bone armatures were also made.

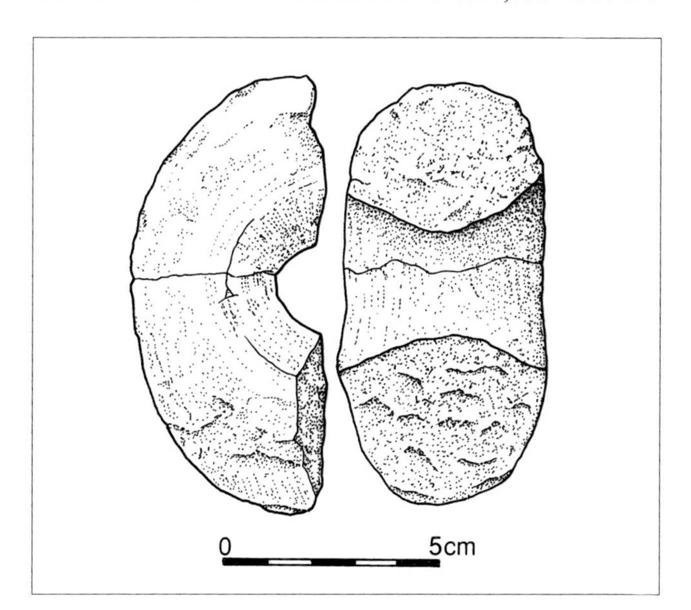

14 Mesolithic 'macehead' (redrawn after Edmonds)

Although the idea may seem rather incongruous from a modern standpoint, we should perhaps also consider the idea that hand-thrown stones could have been used by Britain's Late Upper Palaeolithic and Mesolithic communities in armed conflicts (as they may well have been in later prehistoric periods). Using hand-thrown stones in warfare is certainly not unknown, and for example, Australian aborigines could throw stones very quickly and accurately at their enemies; medieval Swiss mercenaries carried stones in their pockets when going into battle and at the Battle of Kappel in 1531, their formidable stone-throwing is said to have checked an enemy cavalry charge (Chenevix Trench 1972: 10). Although obviously not long-range weapons, hand-thrown stones could certainly inflict serious damage (if not death) when thrown from close range, particularly if they were being thrown en masse by a large group of people.

Interestingly, analysis of 144 skeletons of the Chinchorro people of northern Chile indicates that hand-thrown stones were used in armed conflicts, as healed depressed wounds on skulls and healed facial injuries are likely to have been caused by round, hand-thrown stones (Standen and Arriaza 2000). It has been suggested (*ibid.*: 246) that this skeletal evidence could point to violent mock fights similar to those recorded among the Yaghan Indians of Patagonia, or be related to personal disputes over harvesting grounds or mates. While it appears to be the case that Chinchorro conflict was mostly non-lethal in nature and probably involved low intensity confrontations between males (*ibid.*), it should be pointed out that a young Chinchorro male of 16-17 years had the tip of a quartz projectile point embedded in his spine, which killed him (*ibid.*: 241).

CHAPTER 2

THE NEOLITHIC C.4000-2500 BC

There can be little doubt that our view of Neolithic Britain has been somewhat coloured by an archaeological tendency to portray this hugely important period in Britain's prehistory as essentially peaceful. Thus the farming communities of this time are seen as being preoccupied with tending to their crops and livestock, making stone tools and pottery, and building impressive funerary monuments for their dead. However, it is becoming increasingly evident that if we look closely beneath this peaceful veneer at the archaeological evidence from this period, a rather different picture emerges, as there is compelling evidence that the Neolithic people of Britain were no strangers to warfare.

THE ARRIVAL OF FARMING IN BRITAIN: THE CLASH OF TWO CULTURES?

Firstly, we must consider the possibility that during the Mesolithic/Neolithic transition around 4000 BC, when farming arrived in Britain to replace the hunting and gathering way of life that had lasted for many thousands of years, there was warfare between immigrant farming groups from the continent and Britain's indigenous Mesolithic hunter-gatherer communities. Although the evidence to support this idea is lacking, it is not inconceivable that the arrival of such groups acted as a catalyst – at least in some areas of Britain – for armed conflict between Mesolithic indigenes and Neolithic immigrants. As Dennell (1984:110) has argued, we should be cautious in assuming that all indigenous hunter-gatherers would have welcomed incoming agricultural groups with open arms, as the latter may have posed a threat to their procurement strategies and way of life. Likewise, Spielmann and Eder (1994: 317) have argued that in Europe in general, any incoming Neolithic groups would probably have needed to use a great deal of the Mesolithic land base, not only for the growing of crops, but also for the grazing of livestock.

Of course, such theories are debatable, as the actual impact that the arrival of Neolithic immigrants may have had on the Mesolithic land base could have been negligible, with settlements only consisting of one or two houses in small agricultural clearings (Rowley-Conwy 2004: 96). However, as Peter Rowley-Conwy (*ibid.*) also points out, in time, field size and agricultural production would have increased and substantial areas of coppiced woodland would have been needed for fencing pig [and cattle and sheep] herds, and so, tensions could have arisen between native hunter-gatherer groups and immigrant farming communities as the latter encroached on Mesolithic land. It is also possible that in some cases, it was an irrational fear of people of a different culture that provided the impetus for warfare between hunter-gatherers and immigrant farmers. As Azar Gat (2008: 140) has noted, 'Both among other primates and among humans, field observations and laboratory tests have demonstrated that strangers trigger an initial response of high alarm, suspicion, insecurity, *and* (my emphasis) aggression'.

It should also be borne in mind that Mesolithic hunter-gatherers in Britain may have been tempted to raid immigrant farming settlements for crops and animals, as the stationary nature of these settlements would mean that they would have been vulnerable to such raids, which were probably small-scale, hit and run affairs (*ibid.*: 166-167). Evidence from the ethnographic record lends some support to this idea. For example, early European accounts reveal that the San Bushmen hunter-gatherers of southern Africa raided the livestock of their pastoral and farming neighbours, the Khoikhoi (Hottentots) and the Bantu, and that these raids could spiral into open warfare (Keeley 1996: 133). In the semi-arid regions of southwest USA, Apache and Navaho hunter-gatherers frequently raided the defended pueblo settlements of their agricultural neighbours, mainly for farm animals (Gat 2008: 167-168). These raids could lead to more serious outbreaks of armed conflict; the western Apache of Arizona would form large retaliatory war parties (which could include as many as 200 warriors), that would return to offending settlements to kill as many people as possible if any Apache warriors had been killed in the original raid (Keeley 1996: 135-136).

In addition to ethnographic accounts of warfare between hunter-gatherers and farmers, excavations undertaken at Late LBK (after the German *Linearbandkeramik*), or Linear Pottery Culture sites (*c.*5000 BC) of Darion, Oleye and Longchamps in north-east Belgium (Keely and Cahen 1989) suggest that these sites were fortified border villages which lay on the edge of a 'no-man's-land' that separated immigrant LBK farmers and native hunter-gatherers. The settlements were surrounded by complex enclosures which comprised of deep V-sectioned ditches (over 2m deep), multiple palisades, and baffle gates and at Darion, there may even have been a watch-tower. It is very probable that such features were defensive in nature and that the LBK colonists at these sites did not expect a friendly reception from Mesolithic communities in the region. Although there are no clear signs that the sites were actually attacked by Mesolithic war parties, it is worth noting that five burnt longhouses were found at Oleye and small numbers of Mesolithic projectile points were found both here and at Darion.

However, the possibility that these sites could actually relate to warfare between LBK communities should not be ruled out, because as will be seen below, it is clear from the site of Talheim in Germany, 'that the Linear Pottery colonists also had to fear other farmers like themselves' (Bogucki 1998: 49).

It has to be said, however, that while the ethnographic and archaeological evidence indicates that hunter-gatherers and farmers may well have been involved in deadly clashes elsewhere in Europe, the idea that such warfare may have happened in Britain would not be particularly well supported today. Not only do we have a lack of evidence for its occurrence, but the idea that immigrant communities introduced farming into Britain, which was a favoured theory in the earlier and mid twentieth century, is now decidedly unpopular. As Alison Sheridan (2000: 12) has commented, 'Debates about the Mesolithic-Neolithic transition in Britain have tended to be dominated, in recent years, by a southern English-oriented view that the changes were largely effected by the indigenous gatherer-hunter-fisher groups, selectively adopting resources and ideas from Continental farming communities. Movement of any agricultural settlers has become an interpretative *bête noir*, confined to the wastelands of unfashionable archaeology'. To a large extent, this view still largely holds sway, but while many prehistorians have been happy to consign agricultural farming 'colonists' or 'pioneers' to the scrapheap of prehistory, not all scholars have been willing to accept the general consensus.

Among the dissenters is Anne Tresset (2000: 21) who has pointed out that the earliest finds of cattle, pig and sheep/goat bones from southern British sites date to around the same time (*c.*4000 BC) as those from sites of the Neolithic Chasseen and Michelsburg cultures in the Paris basin. Furthermore, not only are the British animals physically similar to their French counterparts, but as in France, there is an east-west split that can be observed on the British sites, with pigs the dominant species on eastern ones and cattle the more abundant on the western ones. It is also beyond doubt that the domesticated animals found on the southern British sites were imported as the only domesticated animal to be found in Britain before the advent of farming was the domesticated dog (Pryor 2003: 122). Of course, importation does not prove immigration, but as Tresset (2000: 21) rightly notes, it could nevertheless suggest that colonists were an important part of the inception of the Neolithic in southern Britain.

Other evidence that may point to the arrival of Neolithic immigrants in Britain has been highlighted by Alison Sheridan (2000: 6), who has noted the striking

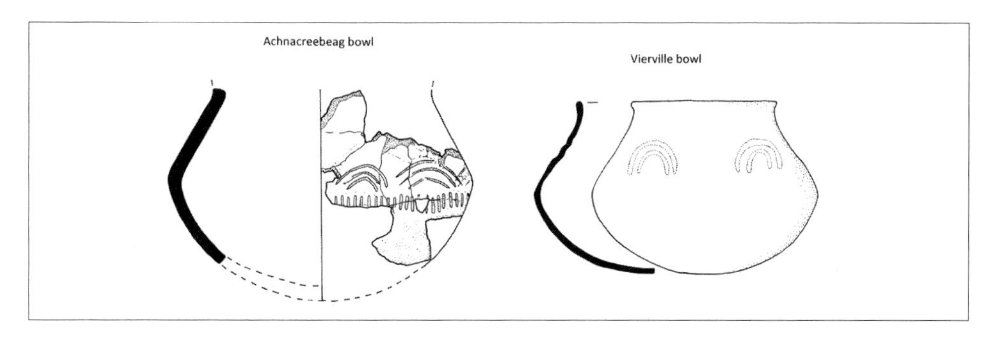

15 Castellic-style bowl found at Achnacreebeag (redrawn after Sheridan)

similarities seen between a finely made Neolithic bowl found in the two phase simple passage grave at Achnacreebeag in Argyll and a Breton, 'Late Castellic' bowl found in a simple passage grave in Normandy (*15*). Sheridan has also noted (*ibid.*: 11) that the architectural sequence of events at the Achnacreebeag tomb is paralleled by Neolithic funerary traditions in the Morbihan area of Brittany, where simple passage graves were juxtaposed onto initial closed chambers. Whether the evidence from Achnacreebeag actually represents the arrival of a north-western French farming community in Scotland around 4000 BC is undoubtedly open to question, but it is an idea that should not be simply discarded.

Other scholars have questioned the prevailing view regarding the arrival of the Neolithic in Britain. For example, Rowley-Conwy (2004: 97) has argued that the archaeological evidence suggests it was probably a rapid process that frequently involved movements of people – though not necessarily in large numbers. Likewise, Rick Schulting and Martin Richards (2002: 177-178) feel that the transition to agriculture on Scotland's west coast was also a rapid event and that it must have involved some level of colonisation.

It is clear that the arrival of farming in Britain around the start of the fourth millennium BC is an extremely complex process that varied from region to region (Noble 2003: 21) and while it is very likely that its Mesolithic inhabitants played a significant part in the transition to agriculture it is also probable that immigrant farming groups were involved to some extent. We may never know whether these farmers engaged in warfare with British Mesolithic communities, as skeletal remains and sites dating to the Mesolithic/Neolithic transition period when such warfare may have taken place are extremely thin on the ground. Therefore, we are undoubtedly straying into the realms of speculation in regard to this idea. Nevertheless, the anthropological and archaeological evidence briefly examined above indicates that we should at least pause for thought before rejecting it. We should also perhaps consider the possibility that Mesolithic communities that had 'become Neolithic' fought with those communities who refused to adopt the new way of life that was centred on agriculture.

Whatever the truth is in regard to the arrival of farming in Britain and the ensuing interaction between farmers and hunter gatherers during the inception of the Neolithic, we will turn now to the question of warfare between Britain's first farming communities.

NEOLITHIC WEAPONRY

Any discussion of the weapons that were used by Britain's first farmers as they strove to kill each other (and also perhaps, some of Britain's final Mesolithic hunter-gatherers) in the armed conflicts that undoubtedly took place during the Neolithic, has to of course include the ubiquitous flint and stone axes that are synonymous with this period in British prehistory (*colour plate 3*).

Undoubtedly, axes were a vitally important practical tool used in mundane everyday tasks such as woodland clearance, carpentry and digging holes. It is also evident that they had a deeper significance that went beyond the ordinary day-to-day concerns of Early Neolithic people. This is indicated by such discoveries as the beautifully made, polished stone axe that was found beside the famous Neolithic wooden trackway, known as the 'Sweet Track' which had run across a bog in the Somerset Levels during the Early Neolithic (dendrochronolgy dated the felling of the timbers used in its construction to the winter of 3807-3806 BC). Subsequent scientific analysis of this axe revealed that it had made a remarkable journey and had travelled many hundreds of miles from a source in the western Alps. How this axe ended up in a bog in Somerset is an ancient mystery, though it is not beyond the bounds of possibility that it was actually carried by someone who came from its area of origin. It is highly unlikely that such a beautiful object, which would have taken a considerable amount of time to make, would have been accidentally lost and the likelihood is that it was deliberately deposited in a ritual act. This theory finds further support when we consider that another fine stone axe, which had come from 'factories' in Sussex, was also found beside the trackway (Pryor 2003: 131).

However, leaving aside the practical and 'spiritual' aspects of Neolithic axes, it is very likely that they were used in Neolithic warfare in Britain, as they would be very effective weapons in close-quarter combat. That they were certainly used in the Neolithic period as lethal weapons that could brutally cut short people's lives is most vividly displayed at the now famous site of Talheim in south-western Germany. Here, excavations revealed a chaotic but well preserved mass of human bones which represented complete skeletons that had been thrown unceremoniously into a pit. The sherds of LBK pottery found among the bones indicated that the grave dated to *c.*5000 BC and subsequent radiocarbon dating of the skeletal material confirmed this (Bogucki 1998: 48). There were 34 individuals represented in the grave; 16 children, 18 adults (many of whom were in their 20s but there were also several people in their 50s and one person over 60. No infants were present (*ibid.*).

The apparent lack of respect accorded to the people in the grave must have alerted the excavation team that they were on to something sinister; the obvious signs of severe trauma that were subsequently discovered on many of the skulls confirmed this. Over half of them showed that many people had received violent blows to the skull and that these blows had produced generally ovate holes that were 2-3cm wide and several centimetres long. Depressions and deformations caused by blunt objects were also visible on some skulls and two showed traces of flint arrowhead impact (*ibid.*). There were also three individuals (2 men and one person of unidentified sex) who appear to have been shot in the back by arrowheads (Vencl 1999: 60) and this suggests that they were trying to escape their would-be killers.

As Peter Boguki (1998: 49) has noted, the shape of the holes and depressions on the Talheim skulls closely corresponds to the cross-sections of the polished stone axes used by the farmers of the Linear Pottery Culture. The true story behind the grim discovery at Talheim remains beyond our reach, but a likely scenario is that an LBK village was raided and overwhelmed by an attacking force and that its inhabitants, young and old, were subsequently massacred by people of their own culture.

Although there is no definite evidence that stone and flint axes were used in Neolithic warfare in Britain, we have much firmer evidence in this regard when we turn to Neolithic archery, as it is clear from both skeletal and site evidence that the bow was used in Neolithic warfare and it is probable that bows were primarily weapons of war. This idea finds support from the fact the bones of wild animals are often only found in small numbers on Neolithic settlement sites (Mercer 1999: 148). Therefore, there appears to be more than a little truth in the words of Humphrey Case (1969: 71), who states: 'Large numbers of arrowheads combined with small evidence for hunting suggests that warfare may have been a seasonal occupation of stably adjusted Neolithic communities in our islands'.

EXAMPLES OF NEOLITHIC BOWS

Unsurprisingly, the wooden bows used by the Neolithic people of Britain are seriously under-represented in the archaeological record. However, we are extremely fortunate to have two well-preserved examples that were found within a mile of each other at Meare Heath and Ashcott in the Somerset Levels (Clark 1963: 55) and which were made around the middle of the fourth millennium BC (Mercer 1999: 147). Although in both cases only half of the bows survived, this allowed their approximate lengths to be estimated and it is evident that the Meare bow had stood at around 190cm, while the Ashcott bow was shorter by about 30cm (Clark 1963: 55-56). Both bows were made from yew, a wood that provides excellent staves for bows because of its slow growth; narrow growth-rings mean that it is particularly resilient to the forces of tension produced when a bow is drawn for firing (Mercer 1999: 145). Nocks for the attachment of bow-strings were present on the surviving ends of both bows, but that on the Meare bow survived in an incomplete state in contrast to the Ashcott bow. Although both bows were similar in height, they were somewhat different in appearance, as the Meare bow was broader in width and remarkably, it still featured remnants of the elaborate ox-hide webbing and binding that appears not only to have acted as a strengthening device, but also as a decorative feature (Clark 1963: 58). The reconstruction of the bow (*16*) gives us a good idea of its original appearance. Experiments have shown that the Meare bow could shoot an arrow over 100m and was the same length as the famous longbows used by medieval English and Welsh archers to such devastating effect at the Battle of Agincourt in 1415 (Pryor 2003: 130). It has been calculated that these latter bows were capable of penetrating plate and chain armour from *c.*50m (*ibid.*). Therefore, as the Meare bow must have contained similar power, in the right hands, it would have been a very deadly weapon.

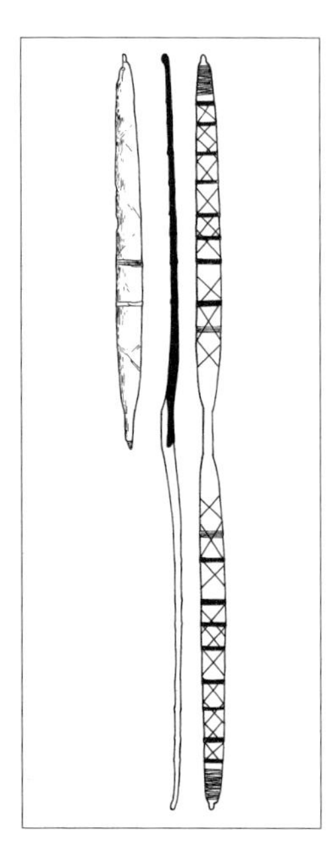

16 The Meare bow as found, and reconstruction (redrawn after Coles & Orme)

Whether the Meare and Ashcott bows were ever actually used as weapons of war can of course never be proved. However, it has been suggested (Coles and Hibbert in Green 1980: 179) that they may have been used in inter-tribal conflicts that flared up over the intercommoning of animals on the Somerset moors. Armed conflicts may also have broken out because of a scarcity of food and land resources during the wet season or over access to the rich grazing areas known as 'hangings', which began to emerge as the flood waters subsided (*ibid.*).

Another Neolithic bow has been found next to the Sweet Track and like the Meare and Ashcott bows only one half of the bow was represented. This suggests that like the former weapons, this bow had been discarded after it had snapped when it was being drawn to fire, although it is possible that these bows were deliberately snapped or 'killed' in acts of ritual deposition. The bow was made from a hazel stave and although it was rather rough and ready and had bark still attached to it in some places (Coles et al 1973: 281), this does not preclude its use as a weapon that could kill people.

An Early Neolithic (4040-3640 BC) bow was also found at Rotten Bottom in the Tweedsmuir hills in Peeblesshire and while it was probably similar in length (*c*.1.7m) to the bows from Somerset, it was a lighter weapon and would have had around half the drawing weight of the Ashcott bow (Mercer 1999: 147).

LEAF-SHAPED ARROWHEADS

For at least a thousand years the standard arrowhead of the British Neolithic was the leaf-shaped arrowhead and these distinctively shaped and finely worked armatures were used throughout the British Isles (Mercer 1999: 148). Although some variation in form can be seen with the kite-shaped and ogival arrowheads, these are probably just variations on a theme and represent differences in raw materials and local traditions of manufacture rather than distinct types of arrowhead (Edmonds 1995: 46). As Roger Mercer (1999: 149) has noted, in both Britain and Ireland, leaf-shaped arrowheads show a remarkable homogeneity in their design and it could be postulated that this 'standardisation' was related to the need to produce an efficient projectile for use in warfare. In fact, Niall Sharples (1991: 48) has argued that 'The leaf shape of the Neolithic arrowhead does not make any sense in regard to hunting and could only be designed for warfare and killing people'.

While leaf-shaped arrowheads are excellent artefacts in terms of their capability to survive the ravages of time, the wooden shafts into which they fitted are obviously not. Nonetheless, there are a number of surviving examples although, not surprisingly, these are in a fragmentary condition. The best known example comes from Blackhillock Bog near Fyvie in Aberdeenshire (*colour plate 4*). This arrowhead is still attached to the partial remains of a wooden shaft that was made from the wood of the Guelder Rose (*Viburnum opulus*), which may have been a favoured wood for the production of Neolithic arrowshafts (Mercer 1999: 147). The shaft is about 23cm long, although it would obviously have been much greater in length originally.

It has been suggested (*ibid.*: 148) that the suitable length for the arrows used with the Meare and Ashcott bows is around 1m (Mercer 1999: 148). Whether the intended target was animal or human, arrowshafts of this length tipped by symmetrical and weighty leaf-shaped arrowheads would certainly have made formidable killing implements when used in tandem with powerful (and not so powerful) Neolithic bows.

OTHER WEAPONRY

As in other periods of prehistory, examples of wooden weapons that might have been used in Neolithic combat are obviously rare, but there have been some intriguing discoveries in this regard. At the important Neolithic site at Ehenside Tarn in Cumbria, two clubs (one of which displays a lattice pattern) and a throwing stick were discovered along with other wooden implements (*17*) (Coles *et. al.* 1978: 8). Wooden clubs and a spear have also been found alongside the Sweet Track, although one of the 'clubs' may actually have been an ard for preparing the ground for Neolithic crops (Coles *et. al.* 1973: 281). The spear, which was 1.85m long, had a sharpened point and a rounded butt and was made from a hazel stave that could have come from a coppiced stand (*ibid.*).

There are a number of other possibilities to bear in mind when considering weapons that may have been used in Neolithic warfare and among these are the antler picks that have been found on many Neolithic sites (*18*). Of course, the prime function of these implements must have been as digging tools. This is borne out by the discovery of antler picks in the ditches of Neolithic ritual and ceremonial monuments and also in Neolithic flint mines. However, if these implements were indeed sometimes used in Neolithic warfare, there can be little doubt that antler picks could cause brutal injuries, particularly if blows were landed on the head or face of an opponent with the protruding branches or tines of the antlers. The pedicles (where the antler had been attached to the head of the stag) would also be capable of fracturing skulls. These ready-made weapons that nature had provided would therefore have been a useful addition to the arsenal of weapons used by Neolithic warriors in face-to-face combat.

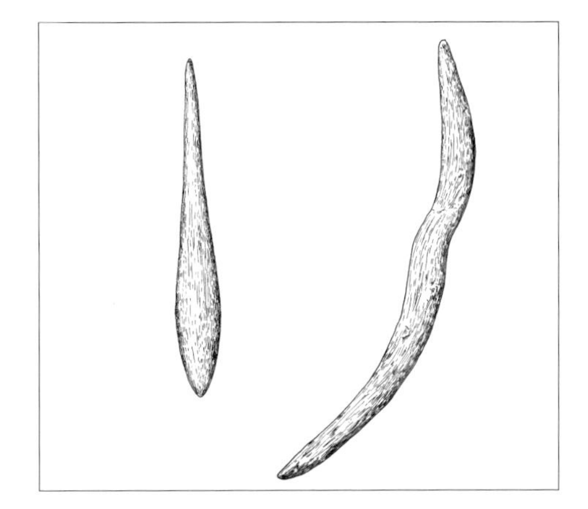

17 Wooden club and throwing stick from Ehenside Tarn, Cumbria (redrawn after Coles et al.)

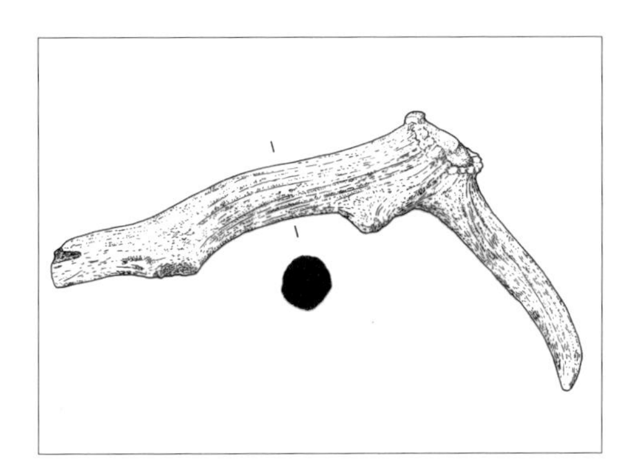

18 Antler pick (redrawn after Wainwright)

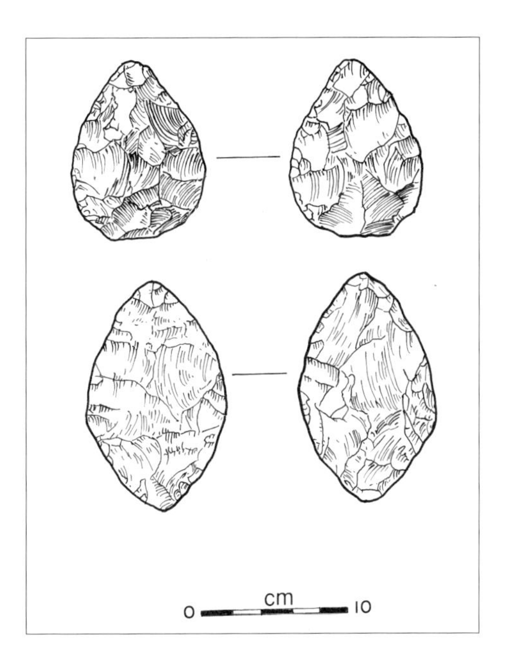

19 Laurel leaves (redrawn after Edmonds)

There are also the somewhat enigmatic artefacts known as 'laurel leaves', which closely resemble leaf-shaped arrowheads. Although they are rather larger in size than the former they were also skilfully made by the bifacial working of flakes of stone or flint (*19*). Because of their shape, it is not surprising that they have sometimes been interpreted as spearheads, although as they are often slightly asymmetrical, this would have impeded their flight (Edmonds 1999: 47). Nevertheless, it is not impossible that they were used as spearheads (*ibid.*) and it could be speculated that they could have been used on shorter stabbing spears employed by Neolithic combatants who fought at close quarters.

As we will see below, some of the Neolithic crania highlighted by Schulting and Wysocki (2005) show small healed depressed fractures which are very similar to sling-stone injuries that have been seen on prehistoric skulls from Peru and the Canary Islands, where slings are known to have been used in tribal warfare (*ibid.*: 125). It is possible that the sling was used in Neolithic warfare, as we know that the sling was used in other countries, such as Greece, during the Neolithic (Vutiropulos 1991: 279-280). However, there is no firm evidence for their use in Britain and sling stones are absent from the places that they would be most expected – such as the fortified sites of Crickley Hill and Hambledon Hill that we know were attacked by hostile forces and where they would have provided a useful defensive partner to bows (Schulting & Wysocki 2005: 126). Intriguingly though, at Gwaenysgor in Flintshire, a great amount of Neolithic settlement debris was discovered in association with an undated hilltop enclosure and thousands of small glacial pebbles were also found gathered together in heaps (Glenn 1914: 247). It is hard not to see these as sling-stone caches for the defence of the site, though it has to be admitted that they could easily relate to a much later phase of the site's history.

FORTIFIED SITES AND THE EVIDENCE FOR WARFARE

The Neolithic causewayed enclosures of the early and mid fourth millennium BC are undoubtedly one of the most enigmatic types of prehistoric monument found in Britain (and indeed Europe). Although they vary in form and size, causewayed enclosures consist of short sections of single or multiple circuits of ditches with undug sections or 'causeways' of earth or rock between them. In some examples, the material quarried from the ditches is used to form continuous or segmented internal banks which correspond to the gaps or causeways in the ditches. The purpose of the causewayed enclosures has been much debated since the early twentieth century. Although it is now widely accepted that no single function can be ascribed to them, the archaeological evidence found within their ditches and interiors indicates that large numbers of people gathered inside causewayed enclosures, probably to participate in rituals and ceremonies of both a religious and secular nature. However, striking evidence found at a handful of sites strongly suggests that in some cases at least, causewayed enclosures played a defensive role and were fortified against attack. As we shall see, it is apparent that these sites ultimately failed in this role.

Hambledon Hill

In the late 1970s and early 1980s Roger Mercer conducted archaeological investigations at the impressive Hambledon Hill in Dorset (*20*) (Mercer 1988). In doing so he provided one of the most revealing pictures of how causewayed enclosures could be places of war as well as peace. The excavations uncovered a number of separate causewayed enclosures which were enclosed by a vast series

20 Hambledon Hill (Marilyn Peddle)

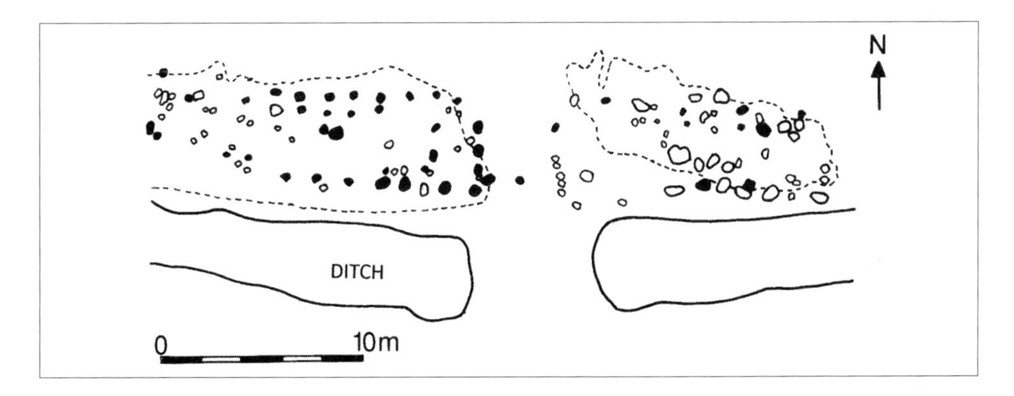

21 Plan of box rampart at Hambledon Hill (redrawn after Mercer)

of earthworks that were reinforced by substantial timbers and cross-beams, to create a box-like rampart that ran a staggering 3000m around the top of Hambledon Hill (21). About 10,000 oak beams would have been required for this rampart and undoubtedly, a considerable labour force allied with highly effective social organisation would have been needed for its construction. There were also three gateways (2.5m wide) through the outworks, whose passages were lined with massive oak posts.

Originally, there were probably three enclosures on the hill, with the third unfortunately all but destroyed by the impressive later Iron Age fort that was built on top of it. The Main Causewayed enclosure was located on the eastern spur of the hill and various items were found in its ditches and interior. These included greenstone axes and gabbroic pottery from Cornwall and two axes of nephrite and jadeite, which probably came from Brittany, or an even more distant source (Mercer 1990: 52).

The smaller Stepleton enclosure lay on the far end of the south-eastern spur of the hill and unlike the main causewayed enclosure, which appears to have been a place more closely associated with ritual, this site probably functioned primarily as a place of occupation. Material that was more domestic and 'industrial' in nature was found here (e.g. flint-knapping debris, animal bones and the remains of an oven or kiln) and large numbers of post-holes also discovered probably marked the sites of former dwellings.

However, while the above material sheds a little welcome light into the darkness of the Neolithic period, evidence was also found that more closely concerns us, as it provides a vivid and remarkable snapshot of warfare during this time. Around 3500 BC, the south-eastern spur of the hill was assaulted by a hostile force and some 200m of the outworks that surrounded the Stepleton enclosure were deliberately set on fire. As Mercer (1988: 104) rightly says, 'The fire can hardly have been accidental in that the whole timber structure was on fire – the oak posts combusting right down into their sockets – and to achieve this effect the rampart would almost certainly have had to be deliberately slighted'.

Even more tellingly, underneath the rubble core of the rampart which had massively collapsed because of the fire, the intact skeletons of two young men were found. It was clear that both men had been killed by archers using leaf-shaped arrowheads, as one was found with an arrowhead in the region of the lungs and it was later discovered that the other had an arrowhead in his throat. The former individual was found with the crushed bones of an infant underneath him and it seems that he was trying to escape with the child in his arms when he was shot in the back and fell into the ditch still carrying this little girl or boy. Such evidence provides us with a sad reminder that as is the case in our own time, the innocent also suffered in the wars fought by our prehistoric predecessors.

Along with the two young men and child found in the ditch, another young male was found on the outer edge of the ditch and he had probably lain here abandoned for some time, as it was evident from the condition of the skeleton that his body had attracted the attention of scavenging animals. Likewise, a partially dismembered skeleton was found in the upper fill of the enclosure ditch, where it may have been dragged by dogs or wolves.

A further two burials were found outside the Stepleton enclosure and it is likely that they also relate to the assault that took place here. One of the burials featured a young male and the fact that he was carefully laid in a pit and covered with scorched chalk rubble from the collapsed rampart perhaps suggests that he was one of the assault force.

Along with the few arrowheads that had obviously been instruments of death, only a few others were found at the Stepleton enclosure and this may seem a little odd at first sight, as the site was obviously attacked by a force of archers. However, it is quite possible, that the evidence found at the Stepleton enclosure only represents one skirmish in a more serious and large scale encounter, with the attackers aiming their main thrust at a possible third causewayed enclosure that lay beneath the later Iron Age hillfort (Mercer 1988: 105).

It has been plausibly suggested (Mercer 1990: 42-43) that the extensive fortifications built around the top of Hambledon Hill may relate to its use as a defensive enclosure, into which large herds of cattle were temporarily driven when enemy raiders intent on rustling cattle arrived in the Stour Valley. Interestingly, analysis of the animal bone assemblage at Hambledon Hill has revealed that cattle were the dominant species and would have been an important source of food (though they seem to have chiefly been dairy animals) for the occupants of Hambledon Hill and the local populace (*ibid.*: 42)

Crickley Hill

Crickley Hill is situated in Gloucestershire along the Cotswold Edge and it forms a prominent landmark (265m) that commands wide-ranging views (*colour plate 5*). Although no victims of warfare have been found at this multi-phase causewayed enclosure, Phillip Dixon's excavations at the site (Dixon 1988) uncovered unequivocal evidence that this site was attacked and destroyed at least once during its history. As at Hambledon Hill, the causewayed enclosure was succeeded by a later Iron Age hillfort.

The first enclosure (which underwent a number of phases of rebuilding) consisted of a double line of widely spaced interrupted ditches that were backed by an inner bank topped by a low palisade. The inner ditch circuit appears to have featured three entrances while the outer had a probable five, and a fence-lined

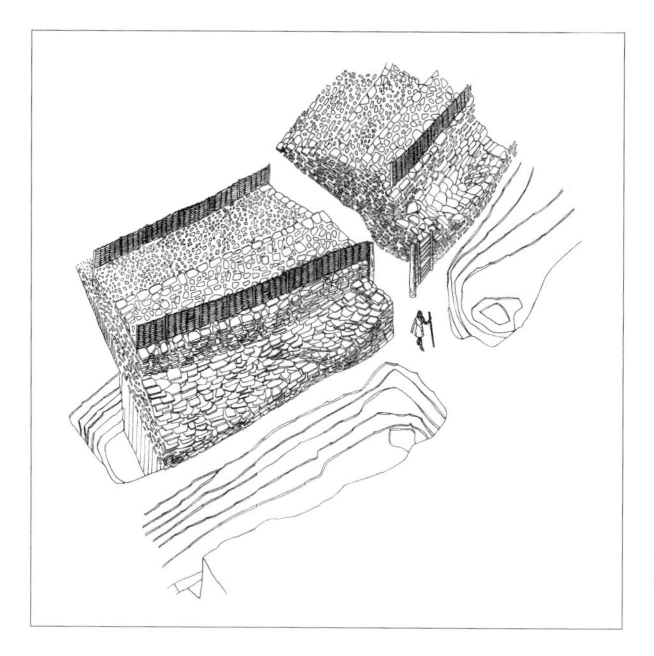

22 Reconstruction of final Neolithic rampart at Crickley Hill (redrawn after Dixon & Borne)

roadway led into the interior. Neolithic pits and post-sockets were discovered inside the enclosure, and although it is not certain what they represent, they probably mark the existence of rectangular Neolithic dwellings. At some point, this earlier enclosure and the buildings inside may have burnt down, as suggested by burnt material thrown into the ditch (Thorpe 2006: 147).

After this possible deliberate destruction of the first enclosure, the site was abandoned for some time before it was subsequently rebuilt. This later phase of the site consisted of a single massive ditch, with two timber-lined gateways and substantial stone-faced rampart that was 10m in width and which again featured a low palisade or breastwork (22) (Mercer 1999: 152). From the gateways, cobbled roads led into the interior where there was a densely packed settlement and an area that may have been set aside for ceremonial activities (*ibid.*).

Evidence found at this later phase of the site is highly suggestive of violent attack and destruction. Over 400 leaf-shaped arrowheads were found clustering around the entrances and arrowheads were also discovered along the cobbled roadways in the interior and along the line of the rampart (23). These arrowheads must surely point to a fierce and concerted assault and it is probable that the archers were marshalled by a fight-leader, or leaders, who directed the attack on the site. The location of the arrowheads could represent a 'softening up' of the defences, in order to demoralise (and kill) the defenders (Miller *et. al.* 1986: 188). It should be pointed out however, that some of them may actually relate

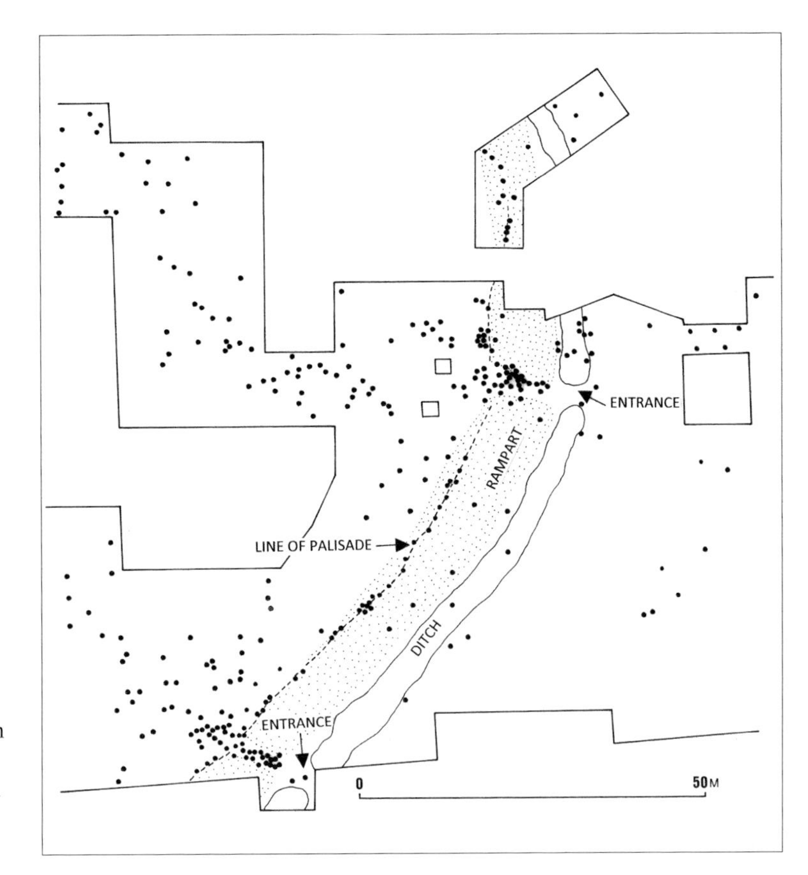

23 Distribution of arrowheads at Crickley Hill (redrawn after Dixon)

to an attack on the earlier enclosure (which it will be recalled showed signs of having been set on fire) and therefore, the amount of arrowheads used in this later attack could be an overestimate (Thorpe 2006: 147).

Inside the enclosure, the remains of a wooden building that had been set on fire were found and among its ruins were over 200 arrowheads (Burl 1987: 38). It is perhaps possible that that this evidence represents a deliberate attack and some sort of Neolithic 'last stand'. However, it is more probable that the building was actually a place where arrowheads for the defence of the site were manufactured, as it is likely that the enclosure at Crickley Hill was designed in the knowledge that assault by an enemy force was a distinct possibility. As Dixon (1988: 82) says: 'The enclosure had quite deliberately been defended against archery attack, and it is highly likely that it was built with this intention, for the low palisade formed no more than a breastwork. The ditches were presumably designed to break up and slow down an assault, and the low bank, or rather platform, would then serve as a killing ground, at point-blank range, against aggressors clambering out of the ditch'.

Hembury

During Dorothy Liddell's excavations at the Iron Age promontory fort at Hembury near Exeter (Liddell 1930, 1931, 1935), she discovered by chance the remains of an earlier causewayed enclosure. Although the evidence discovered here does not point as strongly to warfare as at it does at the above sites, it nevertheless indicates that this site was also attacked.

Along with numerous signs of Neolithic occupation, two sections of a causewayed ditch were discovered, with the longest section near the entrance that led into the fort on the west, and the shorter section to the north-east. A mass of burnt material which included the remains of oak, hazel, ash and also whole oak blocks and sandstone were discovered in the 'inner' ditch and it is likely that this can be related to the remains of a timber palisade or breastwork that had been set on fire (Mercer 1999: 151).

Further evidence of burning came from the remains of a possible house near the entrance causeway. Interestingly, of the 145 arrowheads that were associated with the Neolithic occupation, at least 120 of them (many burnt and broken by fire) came from around the entrance, with a marked concentration of these coming from the area of a likely timber gateway that had also been burnt down (Lidell 1935: 162; Mercer 1991: 33; Mercer 1999: 151).

FURTHER EVIDENCE FOR DEFENCE AND ATTACK AT CAUSEWAYED ENCLOSURES

Along with the above sites, there are a number of other causewayed enclosures which show signs of having been fortified and, in some cases, attacked. For instance, at the Orsett causewayed enclosure in Essex, there may have been a continuous bank or rampart between the two closely spaced outer ditch circuits (a single ditched enclosure that may not be contemporary lay inside these) and the palisade that was set just behind the inner ditch of this double circuit held timber posts that perhaps reached as much as 3m in height (24) (Hedges and Buckley 1978: 236; 238).

On the Isle of Man numerous leaf-shaped arrowheads were found in the ditch of a large D-shaped enclosure at Billown (Darvill 2003: 113-114) and Thorpe's (2006: 148) suggestion that these may be related to an attack on the site is certainly worth considering.

At Donegore Hill in County Antrim, the only known Neolithic causewayed enclosure in Ireland (though there may well be more) consisted of two parallel ditches and an inner palisade surrounding a probable settlement and it is possible that these features were defensive in nature (Waddell 2000: 38-39).

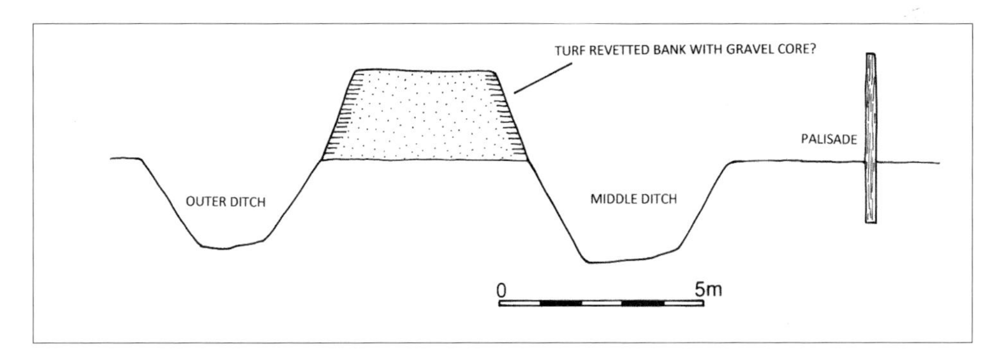

24 'Defences' at Orsett causewayed enclosure (redrawn after Hedges & Buckley)

However, their effectiveness in this respect has been questioned (Thorpe 2006: 148). Also in Northern Ireland, there is the Thornhill site near Derry City, where a Neolithic settlement surrounded by a series of palisaded enclosures has been the subject of limited excavations (Logue 2003). One of the later palisades (Palisade 4) was burnt down, with seven of the 21 arrowheads found at the site (with more probably waiting to be discovered) associated with this episode of destruction (*ibid.*).

Possible evidence of warfare has also been found at the causewayed enclosure that preceded the magnificent Iron Age hillfort of Maiden Castle, in Dorset. Excavations by Mortimer Wheeler in the mid 1930s and Niall Sharples in the mid 1980s discovered concentrations of broken leaf arrowheads in the enclosure ditch. As Niall Sharples (1991: 47; 1991: 255) not unreasonably surmises, this suggests that the Maiden Castle enclosure was attacked by a force of archers and that it may have been violently destroyed as a result.

NEOLITHIC FORTIFICATIONS IN CORNWALL

As has been pointed out (Thorpe 2006: 147), any consideration of warfare in Neolithic Britain must take into account a class of monuments from south-western Britain, which are known as 'tor enclosures'. The most well-known tor enclosure has to be the one found at Carn Brea near Redruth in western Cornwall. The site was investigated in the early 1970s by Roger Mercer, the excavator of Hambledon Hill, and the evidence he uncovered, was again indicative of warfare (Mercer 1981).

The site lay on the eastern summit of the hill and was surrounded by a massive wall of granite blocks (with many weighing 2-5 tonnes) that was integrated with the natural outcrops of granite on the hilltop (25). The wall was some 2m wide at its base and had originally been over 2m high. The discovery of artificial terraces

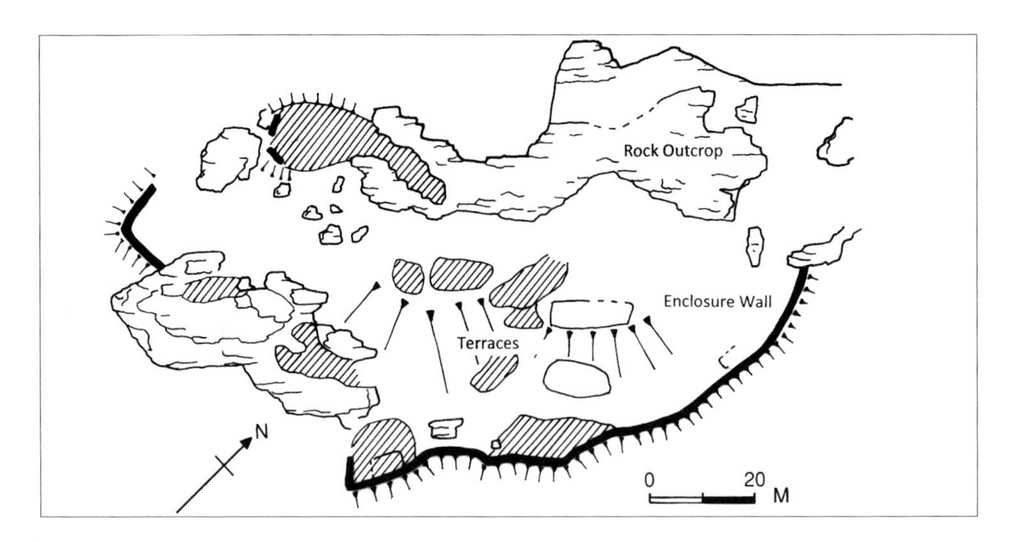

25 Plan of Carn Brea (redrawn after Pollard)

with associated timber buildings and hearths and large quantities of Neolithic flint tools and pottery revealed that this area had once been occupied by a Neolithic settlement or 'village' which was some two acres in size. The enclosure wall actually surrounded an area of over nine acres, some of which could have been used for the cultivation of crops and perhaps also the herding of animals. Traces of an interrupted ditch were also found outside the enclosure wall and it is quite possible that some of the other substantial ramparts that can be found on other parts of the hill belong to the Neolithic settlement on the hill rather than the later Iron Age one. Whatever the true date of these ramparts, other evidence found at the site strongly implied that the eastern summit (and perhaps other parts of the hilltop) had been the scene of a fierce battle between the occupants of Carn Brea and an enemy force, and that the latter had come out the winners in this encounter.

It was clear that a large number of the wooden buildings of the Neolithic village had been burnt and around 800 leaf-shaped arrowheads (many of them burnt white or 'calcined' by intense heat) were recovered from this area and hundreds of other arrowheads have been found on the eastern summit during earlier excavations or as chance finds (*colour plate 6*).

There was also a dense concentration of arrowheads around the area where the entrance into the settlement must have been located (though nineteenth-century dynamiting had unfortunately removed this entrance) and it is possible that the ruinous and collapsed state of the enclosure wall (in which many arrowheads were also found) can be put down to its deliberate slighting by Carn Brea's attackers. The evidence from Carn Brea does not conclusively prove that warfare brought death and destruction to this site during a moment in time in the Neolithic, but as Mercer (1999: 153) states, 'The conclusion that the site's occupation had been terminated by an assault involving massed archers is difficult to resist'.

Some 40km away from Carn Brea is the site of Helman Tor on the edge of Bodmin Moor, which bears striking similarities to Carn Brea. Here too there was a massive boulder-built wall identical in its structure and design to the one at Carn Brea, and this enclosed an area of around 1ha which featured some 19 occupation terraces (Mercer 1986: 53). Small-scale excavations at the site produced similar domestic material to that found at Carn Brea (Mercer 1986: 53) and interestingly, there was a high proportion of arrowheads in the lithic assemblage, with a possible layer of burning also identified (Thorpe 2006: 148). Although the limited work at the site could do no more than provide hints that the settlement at Helman Tor had been attacked, the massive enclosing wall, indicates that at the very least, its occupants were prepared for the worst.

INEVITABLE TARGETS?

It is evident from the artefacts found at the above (and other) Neolithic enclosures that they formed focal points in complex lines of local and wider exchange, along which both raw materials and finished goods passed (Mercer 1999: 153). For example, the community at Carn Brea seems to have been in control of a local stone axe quarry, with 'roughouts' (unfinished axes) being brought to the site for final working and polishing, before they were exchanged for such things as gabbroic pottery from the Lizard peninsula and flint nodules from the Sidmouth/Beer Head area (Mercer 1981: 75). Contacts of a more distant nature were also indicated by the presence of a small number of stone tools found at the site, which probably came from sources in south-west Wales and Great Langdale in the Lake District (*ibid.*).

A similar pattern can be noted from within the 'ceremonial' enclosure at Hambledon Hill, where Gabbroic pottery, Cornish stone axes, quernstones from Devon and two stone axes that probably originated in Brittany (or perhaps from even more distant sources) were deposited in pits (Mercer 1990: 52). At Hembury, we again see gabbroic pottery and greenstone axes from Cornwall, along with querns and rubbing stones from the Sidmouth area, flint and jet from south Dorset and flint possibly from a more distant source (Liddell 1934-35: 162; Mercer 1999: 151). It can be seen then, that enclosures such as these must have been 'wealthy' and important high-status sites that were some way removed from the norm. Therefore, it is possible that they attracted the hostile intentions of 'foreign' communities who sought to usurp the material wealth and the resulting social power that this commanded.

Although the identity of those responsible for the attacks on the communities who occupied these high status Neolithic sites is unknown, it is quite possible that in some instances, their partners in exchange were also their enemies in war. As Schulting and Wysocki (2005: 132) say, 'It has long been a truism in anthropology that groups that fight also exchange, both materials and marriage partners'. Examples of this dichotomous process in action have been noted by Keeley (1996: 122) and can be seen with the warlike Tupinamba of Brazil, who traded with their enemies during regular truces that were put in place to facilitate this trade. Likewise, the truces that were made between the Sioux and Hidatsa Indians along the Upper Missouri only held as long as the two sides were trading at the Hidatsa villages. Once the Sioux were out of sight of the villages, they might kill any Hidatsa they came across and vice versa. Various Eskimo tribes of the Kotzebue Sound region in Alaska would trade goods and generally enjoy themselves at the annual midsummer fair at Sheshalik in July, but by the following November, these very same people could be aiming to wipe each other out in armed conflicts. It is perhaps not unlikely that similar 'fairs' took place at Neolithic causewayed enclosures.

OTHER POSSIBLE NEOLITHIC DEFENDED SITES IN BRITAIN

In addition to Carn Brea and Helman Tor, there are several other tor enclosures in Cornwall that could represent fortified Neolithic settlements. Among these are the Dartmoor sites of Whittor, Dewerstone Rock, Stowe's Pound and Roughtor (26) which have been noted by Silvester (1979).

At Whittor, two low but wide rubble walls surround the site, while at Dewerstone Rock, which sits on a spur above the Rivers Meavy and Plym, there is another double-walled enclosure which has been compared to Carn Brea by Mercer (1981: 191), who notes the massive boulder-built wall joined by granite outcrops that surrounds the site. A later prehistoric hut circle with associated pound can be seen inside the enclosure, but as Silvester (*ibid.*: 188) points out, these may not necessarily be contemporary with the enclosure.

Stowe's Pound consists of a small enclosure surrounded by a massive stone wall that reaches 3m in height and Mercer (1981: 191) has again likened the site to Carn Brea with its smaller 'citadel' enclosure surrounded by a larger outer circuit of walling. The enclosure at Roughtor also features two closely spaced circuit walls, with an additional third stretch of wall on the northern side and two huts of uncertain date. Several hut platforms can also be seen inside the interior of the enclosure.

It has to said however, that although many of the tor enclosures bear some resemble to Carn Brea, both in their design and location, there has been no firm

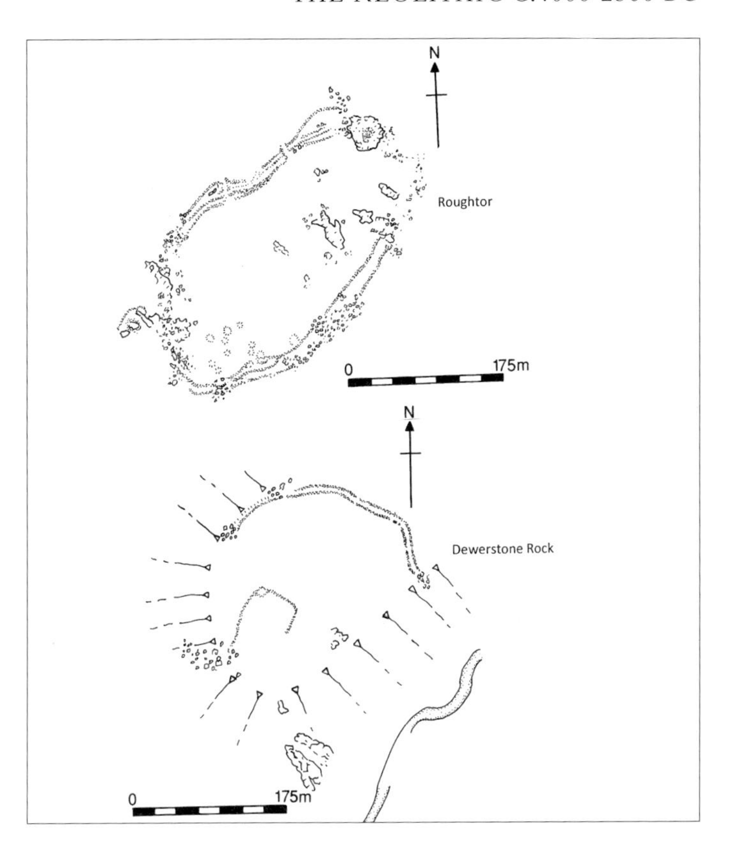

26 Plans of Roughtor and Dewerstone Rock (redrawn after Silvester)

evidence found to support the theory that they were defended Neolithic settlements. However, future excavation at these sites may resolve this uncertainty.

Heading north from Cornwall into Wales, we find the intriguing site of Glegyr Boia near Saint David's in Pembrokeshire. Clegyr Boia is a striking, flat-topped rocky eminence (one of the 'monadnocks' which are found in west Pembrokeshire). Excavations on the summit (Baring-Gould 1902; Williams 1943) revealed that there had been a Neolithic settlement here (perhaps dating to around 3500 BC) and that it may have been defended by a substantial stone rampart and complex gate. Although these defences could date to the Iron Age, apart from a few ambiguous scraps of iron waste, all the artefacts found on site were from the Neolithic.

Equally as fascinating as Clegyr Boia is the site of Gwaenysgor (Glenn 1914), which as mentioned above, produced examples of what seem to be sling-stone ammunition dumps that were gathered together so that the site's defenders would be well prepared for any enemy assault. Gwaenysgor lies some 600ft above the North Welsh coastal town of Prestatyn and consists of a series of stone and earth ramparts that enclose an area of around 4 acres. Although these are now greatly denuded, in their original form they were very substantial walls that in some places reached to 8ft high and 16ft wide. Inside the enclosure, there was an extensive spread of Neolithic material that included several leaf-shaped arrowheads, 'lance-heads' (laurel leaves), fragments of stone axes, a spindle whorl, pottery sherds and the fragmentary remains of cattle, sheep and pig bones. It is also worth noting that a burnt layer, which included some calcined Neolithic arrowheads, was found in the north-western part of the site.

As was the case at Clegyr Boia, the evidence for post-Neolithic activity was lacking at Gwaenysgor. As Glenn (1914: 269-270) says in regard to the occupants of Gwaenysgor: 'They were armed with the bow, lance, sling and axe, and their arrow-heads and lance-tips were leaf-shaped; they were also skilled military engineers, and fortified themselves against a foe expected from the coast'. This statement clearly reflects something of the age in which it was written, but the evidence found at the site perhaps suggests that there may be some truth in it.

Attention should also be drawn to the intriguing Irish site of Ballynagilly in County Tyrone, where the remains of an Early Neolithic burnt wooden building 6.5m long and 6m wide was discovered on a low hill (ApSimon 1971). Interestingly, six leaf-shaped arrowheads were also found at the site, with three inside the house and three just outside. As ApSimon (*ibid.*: 12) says, 'one might wonder whether the arrowheads had been fired at the house before it was burnt down'. Isolated Neolithic hilltop homesteads may actually have been more common than might be supposed, and in this respect we are not helped in our identification of them by hilltop settlements from later periods of prehistory, which have destroyed and confused the picture.

Although unsurprisingly scarce, a number of lone Neolithic buildings have been discovered in Britain, such as the fine 'longhouse' (18m long by 8m wide) discovered under 4m of hillwash at Whitehorse Stone in Kent (*colour plate 7*) (Oxford Archaeology Unit 2000). It is evident that some of these buildings were burnt down; the two impressive timber halls found at the sites of Balbridie and Claish (Barclay *et. al.* 2002), in Scotland and the Neolithic building found at Lismore Fields in Derbyshire (Garton 1987). Like the building at Ballynagilly, these may actually have been Neolithic homesteads that were attacked and destroyed in surprise raids, which was a favoured and unsurprising tactic used in the wars fought by many 'primitive' societies around the world. Of course, it could be that these buildings simply burnt down by accident as the wood and thatch, with which they must have been made, would obviously have been very vulnerable to fire. It is also probable that some of these 'houses' were actually buildings that were connected with the ritual rather than the domestic sphere, as suggested (1991; 9, 25) by Julian Thomas, although these too would still have been very vulnerable to attack by enemy groups.

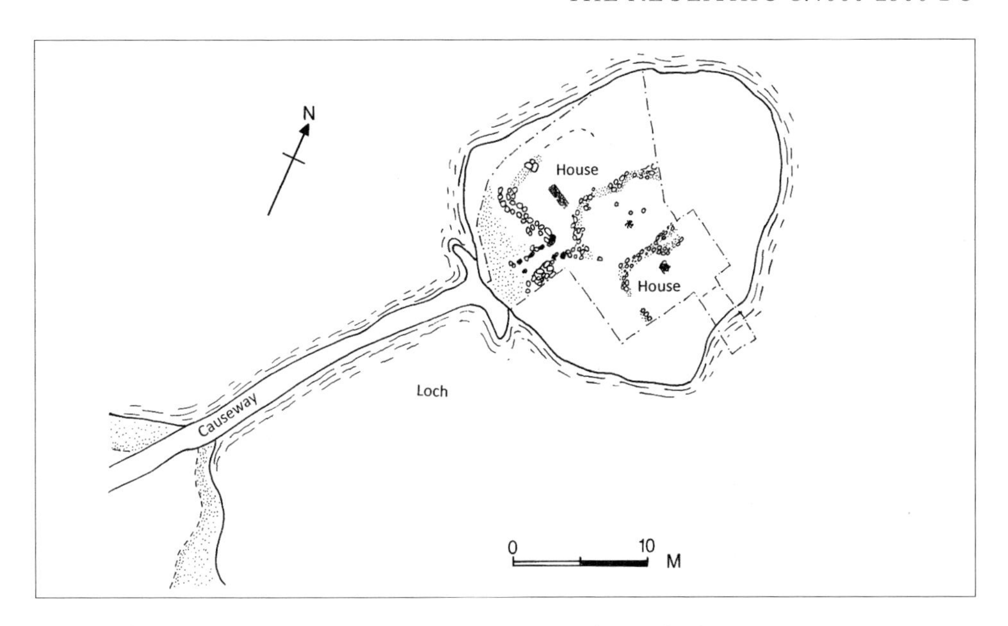

27 Loch Olabhat Neolithic islet settlement (redrawn after Pollard)

There is also the intriguing site of Eilean Domhnuill at Loch Olabhat in Scotland, where an Early Neolithic settlement was located on an islet in the loch. The site showed twelve different phases of occupation and was linked to the loch edge by a stone and timber-lined causeway (27) (Pollard 1997: 12). Although there is no evidence that the site was attacked, could it be that it was fear of this that led Neolithic people to build their dwellings here in this naturally defended location?

SKELETAL EVIDENCE FOR NEOLITHIC WARFARE

As Richard Osgood (1999: 9) has rightly said in respect of the people who lost their lives in prehistoric warfare, 'The most moving evidence for warfare must surely be the victims themselves'. We have already seen some of these victims at Hambledon Hill, with the young men and child who were killed on a dark day now thousands of years in the past. There are several other examples of people who are likely to have lost their lives or been injured in Neolithic warfare.

We will begin our examination of this evidence with the individuals found in various Early Neolithic funerary monuments, who were obviously killed by arrowheads. At Ascott-under-Wychwood long barrow near Oxford, a robust male was discovered in the southernmost chamber with an arrowhead in his third lumbar vertebrae (28 & *colour plate 8*), while in the southern inner chamber, another

28 Arrowhead found in male vertebra at Ascott-under-Wychwood Neolithic tomb (Don Benson)

arrowhead was found lying beneath the ribs of an adult (Selkirk 1971: 8). The position of this arrowhead implies that this second individual was probably also killed by an archer. In regard to the individual with the arrowhead in his vertebrae, Don Benson has made the intriguing suggestion that as the position of the arrow tip suggests an upward trajectory, it is perhaps possible that he was mounted on a horse when he was killed by this weapon (pers. comm.). Although limited, there is some evidence that horses were not unknown in the Early Neolithic, and horse bones have been found in the well known Fussel's Lodge long barrow and at Etton causewayed enclosure (Parker Pearson 1993: 86). Horse bones have also been found in the West Kennet and Winterbourne Stoke (1) long barrows (Burl 1989: 43). However, it should be pointed out that it is not clear whether the bones of Neolithic horses belong to wild or domestic animals. The hinge fracture that was visible on the embedded arrowhead could also indicate that someone tried unsuccessfully to withdraw the arrow from the man's back (Knüsel 2007: 219). It is probable that the arrowhead had cut the main vein of his abdomen, which would have resulted in him bleeding to death (*ibid.*: 218).

At the famous two-phase tomb of Wayland's Smithy above the Vale of the White Horse in Oxfordshire, three leaf-shaped arrowheads were found among the jumbled mass of bones that made up the burial deposit beneath the small oval barrow which had preceded the larger, chambered long barrow. Interestingly, all the arrows had snapped tips (and two had broken butts also) and each was in contact with a pelvis (Atkinson 1965: 130; Whittle 1991a). It was unclear whether these arrowheads were grave goods or were actually responsible for the deaths of the individuals with which they were found (Whittle 1991a: 70). However, recent re-examination of the bones from Wayland's Smithy has revealed that the tip of an arrowhead is still embedded in one of them (Current Archaeology 2007: 20), which strongly suggests that all three arrowheads had killed the individuals.

At Penywyrlod long cairn in the Black Mountains of Wales, the tip of an arrowhead was found embedded in a sternal rib fragment from a young adult. It seems from its superficial penetration of the bone that it was fired from a distance and entered the body at a slight downward angle, perhaps as the victim sought to turn away from the flight of the arrow (Wysocki & Schulting 2000: 599-600).

In the far north of Scotland, at the northern end of Loch Calder, a lozenge-shaped arrowhead was found embedded in the lower thoracic vertebra of an adult placed in the main burial deposit at the circular cairn known as Tulloch of Assery Tomb B (Corcoran 1964-6: 44).

During excavations carried out by the famous General Pitt-Rivers on his vast estate on Cranborne Chase in Dorset, an adult male skeleton with the skeleton of a child at its feet was discovered in the ditch and beneath the two lower ribs of the former there was a leaf-shaped arrowhead (Pitt-Rivers 1898: 63). It is highly likely that this arrowhead had been responsible for the man's death, and one wonders how the child came to be taken prematurely from the world.

In the intriguing multiple burial found during the excavation of an Iron Age settlement at Fengate near Peterborough, an adult male (25-30 years) was found with a leaf-shaped arrowhead between his eighth and ninth ribs (Pryor 1976: 232). The man had been buried with a woman of the same age and an infant of 3-4 years old and a child aged 8-12 years (*ibid.*). Once gain, we are left puzzling over how these individuals lost their lives – did they die through natural causes, or were they also killed violently at the same time as the man?

Across the Irish Sea, the tip of a stone projectile point was found embedded in the right hip of an adult male who had been interred in a Neolithic portal dolmen tomb at Poulnabrone in County Clare (Waddell 2000: 39). Also interred in the tomb were two individuals with healed skull fractures and one also had a fractured rib (Lynch & Ó Donnabháin in Thorpe 2006: 145).

Although not as clear-cut as the above, other Neolithic funerary monuments in Britain have produced evidence that more than likely, indicate that some of the individuals buried in them also met their end at the hands of bow-wielding enemies. Perhaps one of the strongest candidates in this respect is the 'elderly' 50-year-old male (50 would have been a very good age for the standards of the time) whose skeleton was found in the north-east burial chamber of the superb long barrow at West Kennet in Wiltshire. A leaf-shaped arrowhead was found in the region of the man's throat which was 'conceivably the cause of death' (Piggot 1962: 25). He also had a fractured left radius and an abscess cavity at the head of the left humerus, which was probably related to a wound that he had suffered to his shoulder (*ibid.*).

Another possible example of an individual killed by an archer comes from Barrow 13 on Crichel Down in Dorset, where a leaf-shaped arrowhead lay next to the ribs of a skeleton crouched on a bed of flints (Piggot 1954: 49).

Numerous examples of arrowheads have been found in megalithic tombs and long barrows in Britain, sometimes with their tips snapped off at one or both ends (arrowhead tips have also been found in some monuments). For example, arrowheads with snapped tips were found with the sixteen skeletons that lay inside the Harborough Rocks chambered cairn in the Peak District, while at Five Wells Cairn, which lies about 12 miles to the south-west, an arrowhead tip was found among the twelve skeletons buried there (Burl 1981: 95). Several other examples of such finds could be mentioned and doubtless, there are many more that have gone unrecorded. While some of these broken arrowheads probably represent ritually 'killed' grave goods placed with the deceased, it would be flying in the face of reason to suggest that all arrowheads found in Neolithic tombs can be interpreted as such, and at least some (if not many) must have entered these monuments in the corpses of the people that they had killed.

In respect of identifying other possible victims of Neolithic warfare we should undoubtedly consider Schulting and Wysocki's (2005) important and revealing re-assessment of the injuries that can be seen on some of the skulls found by earlier antiquarians and archaeologists in Early Neolithic tombs. Their study looked at some 350 crania from tombs mainly in southern Britain and they discovered that 31 individuals displayed healed and unhealed injuries, with 10 injuries in the former category and 26 in the latter, with a few examples obviously displaying more than one injury.

A striking and clear case of 'perimortem' (occurring at, or around the time of death) injury comes from the skull of a probable adolescent from the Belas Knap tomb. As Sculting and Wysocki (*ibid.*: 111) note, it may or may not be coincidental that the Neolithic settlement at Crickley Hill is not far from this tomb. The adolescent had clearly been hit in the side of the head with considerable force, and this had caused a massive injury to the right side of the skull. What weapon was used to inflict this death-dealing injury is unknown, but it is likely that it was either a stone axe or a wooden or antler club.

A female buried at Belas Knap must also have died from a blow to the top left-hand side of her skull which shows an unhealed, oval depressed facture (*colour plate 9*). It has been suggested (*ibid.*: 125) that the distinctive shape of this injury can perhaps be related to the corner of a polished stone axe, and has been pointed out that some of the skulls found at Talheim display similar injuries.

At the Coldrum tomb in Kent, the skull of a probable adult female shows a clear unhealed injury to the front of the skull (which may have been delivered by a stone axe) and another likely one at the rear. Intriguingly, two fine parallel cutmarks just behind the earhole were also identified and as there are no other cutmarks on the skull, it is quite possible that the woman's ear was cut off and taken as a trophy by the person who killed her (*ibid.*: 129).

Further examples highlighted by Schulting and Wysocki are the adult male skull from an unknown Dorsetshire long barrow, that displays a distinctive 'keyhole' fracture (*colour plate 10*) and an adult female skull from Dinnington long barrow in Yorkshire with a rounded trapezoidal fracture, which again may represent the corner of a stone axe hitting the head with force (*colour plate 11*).

At West Tump long barrow in Gloucestershire, a perforation seen on a skull found here bears a close resemblance to one observed on a pig scapula that had been shot with a flint-tipped arrow during modern archaeological experiments (Smith *et.al.* 2007).

Re-examination of skulls (Smith and Brickley 2007: 23-25) from the well known Boles barrow on the Salisbury Plain military training area, which had been reported by early antiquarians such as William Cunnington and John Thurnam as displaying sword cuts, revealed that several skulls displayed lethal injuries that were probably caused by stone axes. Martin Smith and Megan Brickley (*ibid.*: 26) have made the interesting suggestion that it may only have been people who had been killed in violent circumstances who were interred in Neolithic long barrows. Likewise, Mike Parker Pearson (2005: 26) has suggested that Neolithic tombs and barrows may not just have been monuments for the ancestors, but that they might also have been memorials to significant events such as wars, plagues and famines.

Another possible example of a death caused by a blow to the skull comes from Tulloch of Assery Tomb A, where an adolescent of some 14 years whose skull had been hit with some force was discovered, though whether this blow occurred before or after death is unclear (Corcoran 1964-1966: 32). It is perhaps interesting to recall that an adult killed by an arrow in the back was found in Tulloch of Assery Tomb B, which was located alongside Tomb A.

At the huge Neolithic round barrow of Duggleby Howe in Yorkshire, the skull of a young male was found in a grave shaft at the feet of an old man, and the skull displayed 'a large suspicious-looking circular hole in the left parietal bone' (Mortimer quoted in Loveday 2002: 144). This skull may represent a sacrificial victim killed to accompany his older companion into the 'Otherworld', or alternatively, it could represent a trophy of war (*ibid.*)

As with the unhealed injuries examined by Schulting and Wysocki, the healed examples of cranial trauma are found on both male and female skulls and these take the form of small depressions with rounded edges. Notable examples are the male skull from Fussell's Lodge, which displays three healed depressed fractures in a row and the fractures that can be seen on a female skull found at Dinnington and a possible male skull from Norton Bavant (*colour plates 12 & 13*). At Hambledon Hill, two skulls (one male, one of uncertain sex) with healed injuries have also been recorded (Roberts and Cox in Thorpe 2006: 145).

It could be that some of these healed cranial injuries were caused by sling-stones, although they could just as easily have been caused by such things as the tines on antler clubs, or glancing blows from arrows and wooden-tipped spears (Schulting & Wysocki 2005: 127). It is also possible that some injuries relate to ritualised non-lethal combat, such as the club fights observed among Aboriginal males in Australia (Ostendorf Smith 2003: 314) or the Chinchorro 'stone-fights' which we encountered in the previous chapter.

There are many examples of isolated skulls found at other causewayed enclosures and other Neolithic funerary and ceremonial monuments and it is possible that some are related to warfare, rather than ritual acts of deposition, which is often the preferred explanation. As Aubrey Burl (1981: 7) has said in regard to the skulls found in the ditches of the famous causewayed enclosure at Windmill Hill in Wiltshire, 'They could have been...trophies of battles fought by people who had not enjoyed such a peaceful existence as some archaeologists have believed'. Likewise, Stuart Piggot (1954: 47) also suggested that the abundance of skulls on Early Neolithic sites pointed to headhunting during this time. In fact, there have been a number of skulls found at Neolithic sites with neck vertebrae still attached, which indicates that they were removed from bodies that were still in a fleshed state.

At Bridlington in Yorkshire, the partially cremated remains of a skull with an atlas vertebra still attached were found in a pit during the excavation of a medieval post mill and intriguingly, also discovered in the pit was a Neolithic flint axe (Earnshaw 1973: 22-23). It is not unlikely that it was this axe which removed the head from the person's shoulders. Therefore, could we again be looking at an example of a trophy taken in battle by a Neolithic warrior, later to buried in the ground as an act connected with the worship of unknown subterranean gods?

Another example of what must originally have been a decapitated head was discovered by the well known nineteenth-century antiquarian John Thurnam, at the previously mentioned Boles Barrow, where there was a skull with an attached neck vertebra that had been cut in two (Piggot 1954: 57). Likewise, in the Chute 1 oval barrow (Wilts.), there was an intriguing group of skulls arranged in a circle and one of these still had three vertebrae attached (Passmore 1942: 100).

During excavations at the Staines causewayed enclosure in Surrey, two skulls were found in the outer ditch. It is reported (Robertson-Mackay 1987: 38) that one had healed wounds and that its owner was later violently killed by being struck repeatedly on the head, before he was decapitated. It may also be worth noting here that along with the skulls found at Staines there were other skeletal remains (a forearm and lower jaw) found in the ditches, which also contained occupation refuse such as bones from cattle and pigs, and Early Neolithic pottery (Roberstson-Mackay 1987: 36). As pointed out (Robertson-Mackay 1987: 38), this juxtaposition of human material with occupation refuse suggests that human remains had been lying around on the site, before they were interred in the ditches.

Similar patterns of deposition have been noted at other causewayed enclosures and while it is likely that for the most part, these represents ancestral (or other) rituals, it might be that in some cases, other isolated body parts such as arms and jaws also represent trophies that had been taken from slain enemies (Schulting & Wysocki 2005: 129). It is also not beyond the realms of possibility that some skeletal material found at causewayed enclosures, which bears no obvious signs of trauma, actually represents individuals who had been killed in attacks on enclosures, only later to be buried by friendly (or unfriendly) hands in their ditches some time after they had been lying exposed on the surface.

In addition to the cranial evidence, we have other unhealed and healed trauma present on Neolithic skeletal material in funerary contexts and again, some of this may point to death and injuries relating to Neolithic armed conflicts. In Scotland, a cist-grave discovered underneath a Neolithic cairn near the Glenquickan stone circle contained the skeleton of a large man whose arm had been almost severed by a greenstone axe, a fragment of which was still embedded in the bone (Burl 1987: 112).

Among other examples of healed trauma are several forearm fractures, such as the one seen on the left forearm of a person aged around 50 years who was found in a Neolithic long barrow during excavations at the well known Neolithic and Bronze Age monument complex at Barrow Hills in Oxfordshire (Barclay and Halpin 1999: 29).

Underneath a Bronze Age bell barrow near Dunstable, the primary Neolithic burial was a female whose left forearm showed a healed fracture near her wrist and the excavators plausibly suggested that this had occurred when she had fended off a blow from a club (Dunning and Wheeler 1931: 196). Of course 'parry fractures' on the forearms of Neolithic individuals may relate to accidents, but nevertheless they are still often viewed as evidence of interpersonal violence (Schulting and Wysocki 2005: 123, 124). It is possible that some of this violence took place within the larger context of armed combat between opposing forces that were intent on doing more to each other than just breaking arms. A similar argument could be proposed for the healed rib fractures that have also been noted in some Neolithic burials, although it is probably more likely that these represent accidental injuries.

THE LATE NEOLITHIC: A TIME OF PEACE?

In contrast to the Early and Middle Neolithic, we have no definitive evidence for warfare from the Late Neolithic (*c*.3200-2500 BC). However, it is improbable that warfare was completely unknown during this lengthy period and the lack of skeletal evidence for violent trauma could be due to a general scarcity of complete bodies from the burial record of the Late Neolithic. Also, although as mentioned above we do not have any conclusive evidence for fortified sites during this time, there are possible examples that should perhaps be taken into consideration in this respect and one such site is Castell Bryn Gwyn on Anglesey. In its original form, this site consisted of a circular area some 130ft in diameter that was surrounded by a external ditch (7ft deep and 20ft or more wide) and an inner, boulder-cored bank (25ft wide and perhaps 6ft high) which were pierced by a very narrow causeway on the south-western side (Lynch 1970: 65). Within the interior, no signs of ritual activity were found, though there was a scatter of occupation debris that included large sherds of Fengate Ware pottery. Fengate Ware was produced in southern England in the last quarter of the fourth millennium BC (Pollard 1997: 18) and thus it is conceivable, as Lynch (*ibid*.: 66) suggests, that the ditch and 'bank' (this would have more resembled a rampart) at Castell Bryn Gwyn represent defences built around a settlement which may have been populated by an incoming group of 'English' colonists in the Late Neolithic.

Another intriguing site is Meldon Bridge in Peeblesshire, Scotland. Here a massive timber wall, 600m long and reaching 4m in height, was constructed on a promontory between two rivers, *c*.3000-2500 BC (Speak & Burgess 1999). Although the earlier idea that Meldon Bridge palisade represents a defended settlement is no longer favoured and it is felt that is more likely to be a ceremonial enclosure (Thorpe 2006: 151), the possibility remains that the timber palisade was indeed intended as a defence against any would-be attackers. This is perhaps indicated by the fact that the site was built in a naturally defensible position (which the Romans also took advantage of, as shown by their later construction of temporary marching camps at Meldon Bridge), but also by the possible existence of a double gateway, guard tower and hornwork at the site's entrance (Speak & Burgess 1999: 108).

It may be that warfare during the Late Neolithic is harder to discern because it changed in nature. During this time, the leaf-shaped arrowhead of the earlier

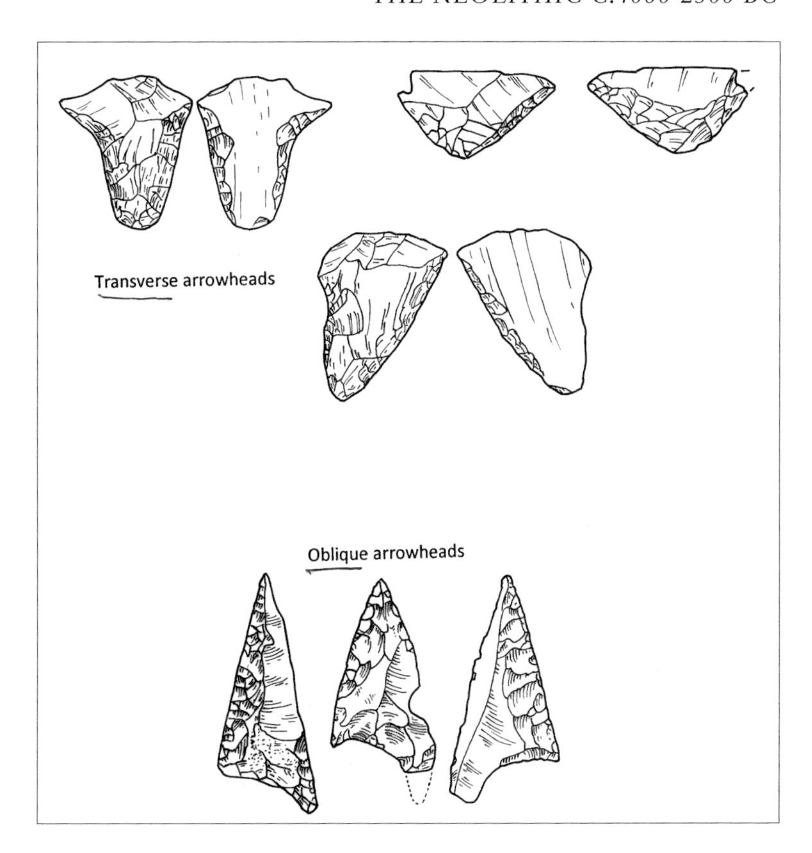

29 Late Neolithic 'petit-tranchet derivative' arrowheads (redrawn after Edmonds & Curwen)

Neolithic was replaced by the 'petit-tranchet derivative' arrowhead that came in a wide variety of transverse, chisel and oblique forms (29). These arrowheads have often been seen as suitable weapons for hunting wildfowl and other small game (Edmonds & Thomas 1987: 193), although it has been argued (Clark in Edmonds & Thomas: *ibid.*) that they would not have been particularly suitable in this respect, as they would break small bones and lead to blood-soaked feathers. It could be then, that they were used in ritualised combat or in cattle raids or feuds and that their primary purpose was to wound and disable opponents, but not kill them (*ibid.*). However, we should remember that such arrowheads would still have been quite capable of killing people and the find of a transverse arrowhead embedded in the vertebra of an individual found in a Late Neolithic rock-cut tomb in France, bears witness to this (Clark 1963: 81-82; Edmonds and Thomas 1987: 193). Therefore, it is quite possible that some of the small number of Late Neolithic arrowheads found in association with burials actually represent the cause of death of the individuals with whom they were found.

Although as mentioned above, the evidence for violent trauma is lacking from the skeletal record of the Late Neolithic, a discovery from the remarkable Late

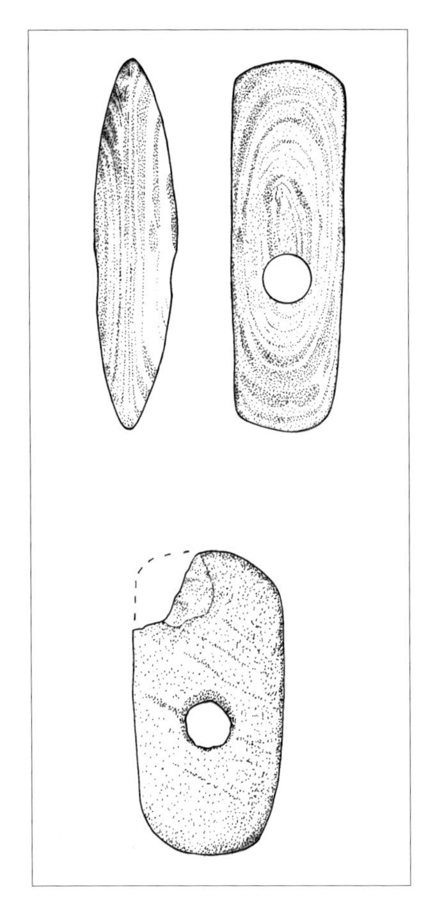

30 Late Neolithic maceheads (redrawn after Edmonds)

Neolithic settlement, at Durrington Walls (currently being excavated by the Stonehenge Riverside project) strongly suggests that an individual was attacked and killed by an archer. In a pit cut into the corner of the north-west of House 457, a human femur with two peri-mortem impact injuries on opposite sides of the bone was deposited at the end of the Late Neolithic *c.*2500 BC. A radiocarbon date taken from the bone was older than the pit, which could either suggest that the bone was 'curated' before burial or that old carbon had been ingested by the individual in their food (Mike Parker Pearson pers. comm.).

Finally, it has also been suggested (Edmonds 1995: 110) that although some of the finely made maceheads (*30*) of the Late Neolithic must have been made purely for display, many could have been used as actual weapons in armed conflicts between separate communities or kin groups.

CHAPTER 3

THE BEAKER PERIOD AND EARLY BRONZE AGE C.2500-1500 BC

Whether one favours the term 'Late Neolithic', 'Beaker Period', 'Early Bronze Age' or 'Copper Age' (see Needham *et. al.* 2008a for a recent discussion on the British 'Copper Age') the 500 or so years that run from *c.*2500-2000 BC undoubtedly represent one of the most fascinating and exciting phases of British prehistory. It was during this time that metals first began to be mined and worked, distinctive pots known as 'Beakers' were made and there was a marked shift from the collective burial rite seen in Neolithic tombs to an emphasis on individual burial underneath low burial mounds and in flat graves. Alongside Beakers and other artefacts, striking types of weaponry not seen before were also sometimes placed in male Beaker graves. Given the predilection of earlier archaeologists to view the appearance of novel items of material culture as a sign of immigration, it is hardly surprising that the Beaker 'package' in Britain was seen to represent the arrival of a new and warlike people from the continent. However, the idea of a Beaker 'invasion' is now rightly treated with scant respect, and today, the concept is largely defunct, with few if any, archaeologists continuing to believe in it. Nevertheless, evidence is increasing that there was small-scale immigration of 'Beaker Folk' (a term that is not particularly *de rigueur* amongst archaeologists these days) into Britain from the continent, and more importantly, 'the impression ... remains of the Late Neolithic as a period of conflict, due to the symbolism of the artefact repertoire' (Thorpe 2006: 151). It is to this repertoire that we will now turn, beginning with the archery equipment found in Beaker burials.

Alongside Beakers and other items, a characteristic feature of the pan-European Beaker phenomenon is the barbed and tanged flint arrowheads and perforated stone plates that accompany male burials (though not always together). The traditional view regarding the latter is that they are archer's wristguards or 'bracers', designed to protect the wrist from the recoil of the bowstring. However, it is evident that in many Beaker graves bracers are located next to the outer forearm of the deceased. Therefore it seems probable that the majority were

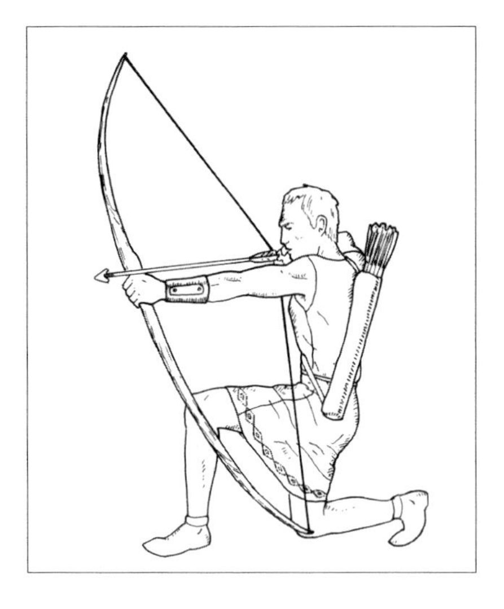

31 Beaker archer wearing bracer in ornamental position (redrawn after Fokkens *et. al.*)

probably mounted on leather wristlets with an ornamental position favoured over a functional position on the inner arm (*31*) (Fokkens *et. al.*: 2008: 116). Evidence for the mounting of bracers on material of some kind has in fact been found on bracers from Britain, with examples from Beaker burials at Driffield (Yorkshire), Borrowstone (Aberdeenshire) and Culduthel Mains (Inverness-shire) featuring bronze rivets still in their holes (Smith 2006: 11). In some cases, it appears that bracers were held in place with organic bindings, such as the fine example found at Hemp Knoll in Wiltshire which still clearly shows the staining left by these bindings (*ibid.*). It is interesting to note that the rivets of the bracers from Culduthel Mains and Driffield were capped with gold, while a fine green stone bracer from the primary Burial at Barnack in Cambridgeshire features nine domed sheet gold caps at either end (*32*). However, unlike its Scottish and Yorkshire counterparts, no traces of rivets were found on the Barnack bracer and so it is hard to see how it could have ever been attached to a wristlet made of leather or some other material (Donaldson 1977: 215). The likelihood then, is that the Barnack bracer was made as a ceremonial item to be placed with the deceased in his grave. Joan Taylor (1994: 48) has made the interesting suggestion that the bracers with gold caps may demote the top rank of the Beaker 'archer class' and that one of their roles may have been to protect the metal ores that were extracted by Beaker mining 'engineers' (pers. comm.).

One of the finest examples of the Beaker burial practice of placing barbed and tanged arrowheads and bracers with the deceased comes from the grave of the now famous 'Amesbury Archer'(*British Archaeology* 2003). The grave (*colour plate 14*) was discovered in early summer 2002, during an excavation by Wessex archaeology in advance of the building of a new school at Amesbury,

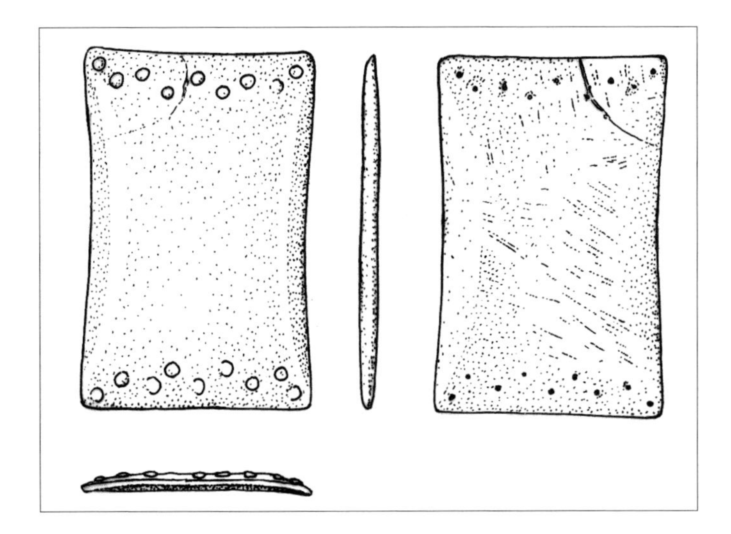

32 Barnack bracer (redrawn after Donaldson)

on Boscombe Down in Wiltshire. It provides us with the most richly furnished Beaker burial yet found in Britain and continental Europe. In the grave lay the skeleton of a 35-45 year old male, buried at some point between 2400-2200 BC, who was well prepared for the afterlife, as nearly 100 artefacts accompanied him in death. Although the majority of these were flint tools (33), the burial was nevertheless unique in terms of the quantity of the more impressive objects that were discovered. Among the artefacts buried with the archer were two sandstone wristguards (*colour plate 15*) with a black/grey one located on his outer forearm (Andrew Fitzpatrick pers. comm.). Fifteen barbed and tanged arrowheads (*colour plate 16*) were also found and these may represent the remains of a quiver of arrows scattered over the man as he lay in his grave. Other grave goods included five early Beakers (*colour plate 17*), three copper daggers made from continental ores (*colour plate 18*), four boar's tusks, a bone pin, two gold basket earrings or hair tresses and a cushion stone, which was almost certainly used for making small metal objects (34). The stunning array of artefacts buried with the Amesbury Archer suggests that he was a person of considerable status and the fact that he was buried around the same time as the building of the famous monument led the press to give him the grand title of 'King of Stonehenge'. Whatever the reality of the Amesbury Archer's relationship to Stonehenge, a surprise lay in store, as oxygen isotope analysis carried out on his teeth revealed that as a child he had lived in Central Europe, probably near the Alps.

Another unexpected discovery was a further burial discovered about 5m from the Archer's grave (35). This contained the skeleton of a man aged 20-25 years who had also been buried with a set of gold earrings/hair tresses similar to the

33 Flint knives and scrapers found with the Amesbury Archer (Wessex Archaeology)

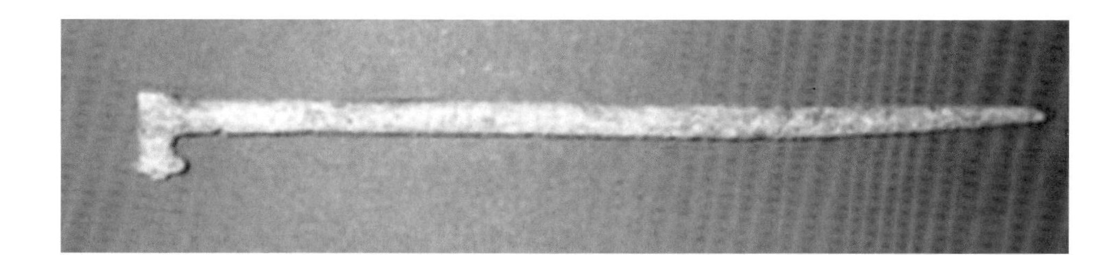

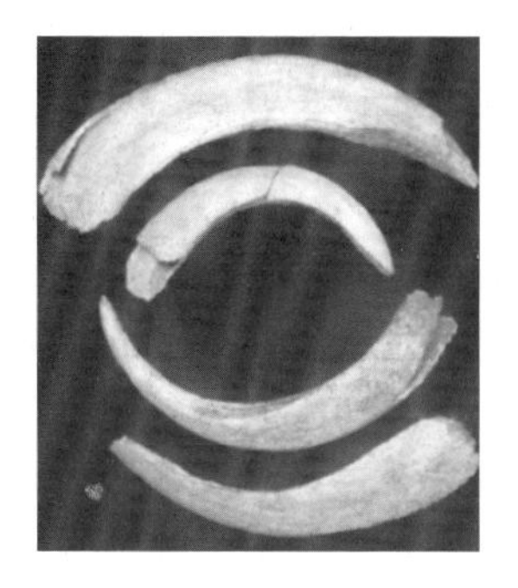

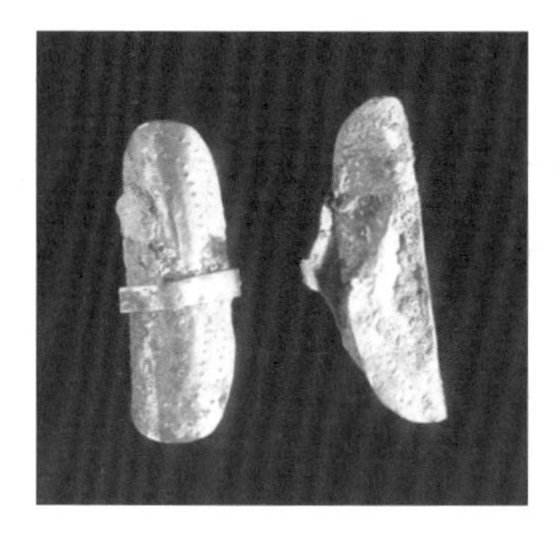

34 Grave goods found with the Amesbury Archer (Wessex Archaeology)

Above Left: 35 The Amesbury Archer's companion (Wessex Archaeology)

Above Right: 36 The Boscombe Bowmen's grave (Wessex Archaeology)

37 Barbed and tanged arrowheads found with the bowmen (Wessex Archaeology)

Archer's. Of even more interest is the fact that analysis of the skeleton revealed that like the Archer, this young man had an unusual bone in his instep and thus it is likely that they were related – it is even possible that they were father and son.

Further examples of Beaker immigrants include the so-called 'Boscombe Bowmen' who were found in an unusual multiple grave on Boscombe Down (*36* & *37*). They could have been born in Wales, but it is perhaps more likely that they originated in Brittany (Sheridan 2008: 27). A Beaker grave recently discovered in the Kilmartin Valley may mark the final resting place of an immigrant from the Netherlands who arrived in Scotland between 2570-2280 BC (*British Archaeology* 2008).

While it is clear that archery was an important male activity among the Beaker communities of Europe and that archers were ascribed some level of status, archaeological opinion is unsurprisingly divided as to whether they were hunters or warriors. Among those in the former camp is Roger Mercer, who feels that it is unlikely that the archery equipment in Beaker burials symbolises the status of a prestigious group of warriors but rather, it is more probable that it signifies a hunting ritual carried out by an elite within Beaker society (2006:128, 133). Possible examples of this practice include the pit which contained the skeleton of an aurochs with six barbed and tanged arrowheads between its ribs, discovered during the building of Terminal 4 at Heathrow, and the young aurochs buried among a number of Beaker ritual and ceremonial monuments discovered on Boscombe Down (*ibid.*: 132).

Whilst it is quite possible that Beaker archers did carry out specialised hunting forays, that were connected to ideas about male status and the 'Other-World', or the evidence perhaps indicates that it is more likely that the primary role of Beaker archers lay in warfare. In this respect, Edmonds (1995: 143) has said: 'Together with an increase in the number of burials where death or injury seems to have been caused by archers, this rise to dominance of a sturdy piercing arrowhead may be of some significance'. It has also been suggested (Fokkens *et. al.* 2008: 123) that an almost paradoxical situation exists during the Beaker period, with archery equipment becoming prominent in the archaeological record in a time when farming was the main means of subsistence. Thus rather than revealing the existence of high-status male hunters, it has been plausibly argued (*ibid.*: 124) that as was the case in a number of non-state societies, this equipment reflects a martial ideology that is symbolic of the important role that warriors held in Beaker society.

BOWS, BATTLE-AXES AND DAGGERS

Many small crescent-shaped pendants (*38*) have been found in male Beaker graves (Piggot 1971) on the continent and it is probable that these are representations of short, but powerful composite bows similar to those used by such peoples as the ancient Scythians, Egyptians and Assyrians. These bows were made from wood, horn and sinew and measured about 1m in length. Some of the Beaker pendants appear to show nocks for the attachment of strings and bindings to strengthen the staves, which lends support to the theory that they are miniature versions of composite bows. Examples of bows from the latter half of the third millennium BC are lacking in Britain and whether such bows were also used by Beaker archers in Britain may never be known, though a possible bone pendant representing a composite bow was discovered with the Barnack bracer mentioned previously (Donaldson 1977: 215; Kinnes in Taylor 1994: 47). However, it has to be said that in appearance it does not particularly resemble a composite bow (*39*). More interesting in this respect is the dark, curved stain of earth (about 1m in length) found close to five barbed and tanged arrowheads underneath a round barrow in Yorkshire and the traces of wood, sinew and leather discovered in a Beaker burial in Aberdeenshire (Burl 1989: 109).

In addition to the archery equipment, other distinctive weaponry in the form of battle-axes (*40*) and – as have seen with the Amesbury Archer – copper and bronze daggers with a tang for the attachment of the hilt, also sometimes appear in male Beaker graves (*41*). Although firm evidence for the use of battle-axes is lacking in the skeletal record, the design and weight of these artefacts suggests that they were not designed for utilitarian purposes; rather, it implies that they were essentially weapons that would have been suitable for use in close-quarter combat, where they would be particularly effective for dealing out brutal blows to the head and upper body (Mercer 2006: 133).

However, in contrast to the battle-axes, copper and bronze daggers may not have been weapons that were used by combatants in face to face fighting.

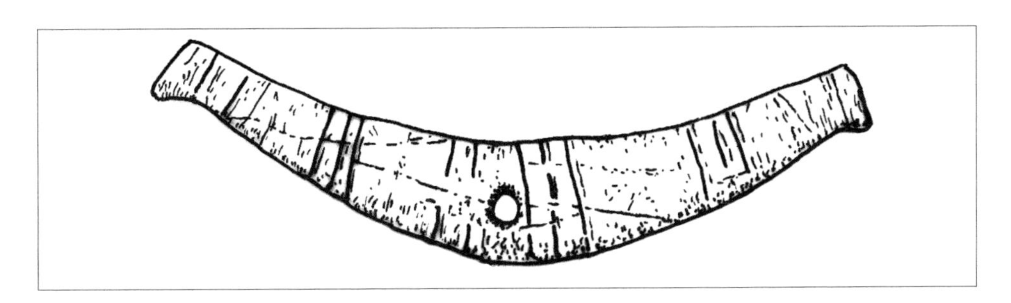

38 Beaker bow pendant from Moravia (redrawn after Piggot)

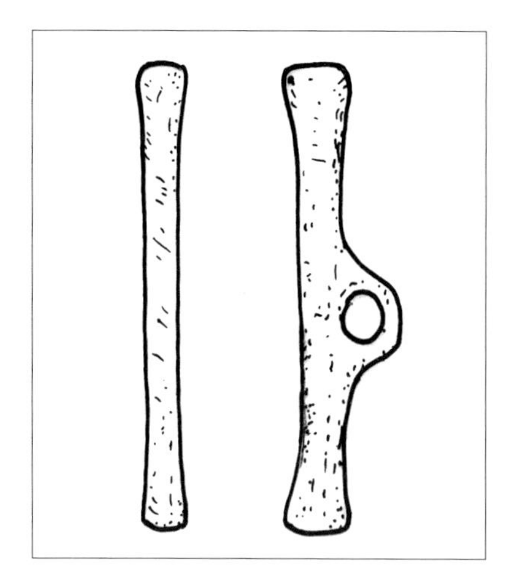

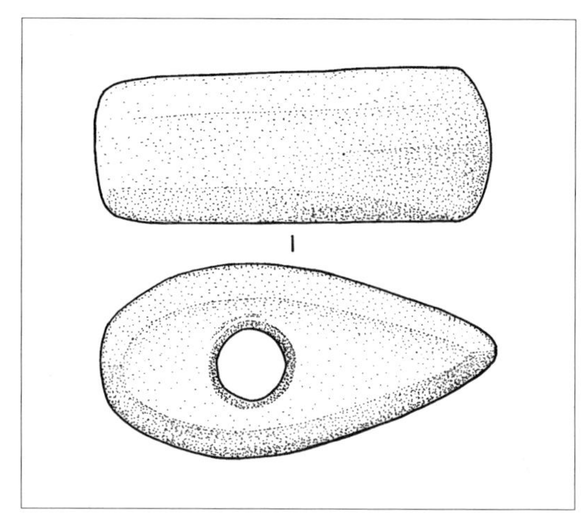

Above Left: 39 Beaker pendant from Barnack (redrawn after Donadlson)

Above Right: 40 Beaker battleaxe (redrawn after dean)

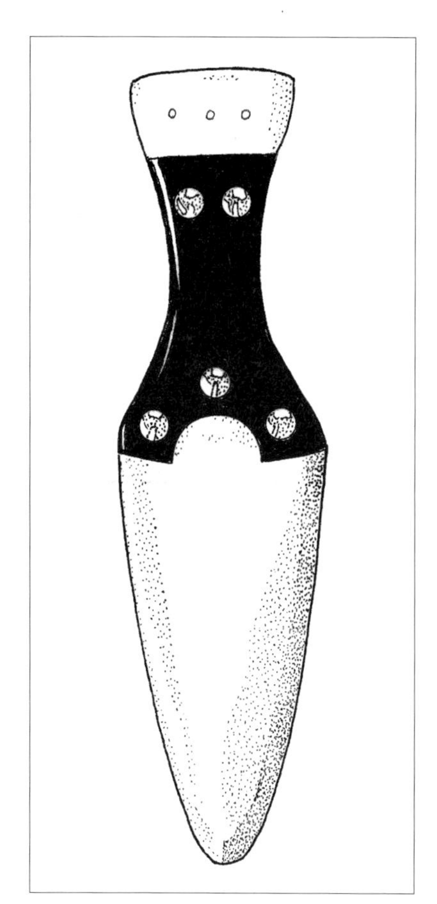

41 Beaker dagger from Ashgrove, Fife

It has been argued (*ibid.*: 124) that their rounded points, triangular form and the general frailty of their construction would make them ineffective for delivering devastating puncture wounds and that as their edges often show signs of whetting, this suggests that it is more probable that they functioned as multi-functional knives, probably used in ritual and ceremonial contexts (*ibid.*: 125). Although Beaker daggers may not have been particularly robust, and it does seems probable that they were multi-purpose implements, they could still have been used as both cutting and stabbing weapons and would undoubtedly have caused lethal wounds to unprotected parts of the body. Humphrey Case (in Mercer 2006: 129) has suggested that Beaker 'knives' were used to cut the throats of wild animals brought down by Beaker archers in the hunt and that their blood was drunk from the Beakers. An alternative (and equally speculative view) is that Beaker daggers were used to administer the *coup de grace* to wounded enemies brought down by Beaker bows. The feared English and Welsh longbow units that formed the backbone of medieval English armies employed a similar tactic on battlefields, where they would use their small and narrow daggers (nicknamed 'ballock' daggers because of the twin spherical grips either side of the blade, which gave them a resemblance to an obvious part of the male body) to stab wounded enemies through helmet visors or unprotected parts of the body. Perhaps though, we would be best to view Beaker daggers as 'one-chance' weapons that may have been used in very close combat, with the status and threat that they signified being of equal or more weight than their actual importance as killing implements (Osgood 2000: 23; Harding 2000: 309).

OTHER WEAPONRY

It is probable that as in other prehistoric periods, wooden clubs were used in combat during the Beaker period, and in this respect, attention should be drawn to the grave discovered at the multi-period ritual site at Cairnpapple Hill, West Lothian (Piggot 1948). In addition to the two crushed Beakers, and the possible remnants of a wooden mask placed over the face of the deceased, the carbonised remains of what had probably been a massive wooden club lay along the north side of this rock-cut grave (42).

During the Beaker period, the first metal axes appeared in Britain, with simple flat copper axes (43) followed by bronze examples around 2200 BC, when British smiths learnt the art of alloying copper with tin to make the more durable material of bronze. A fine example of a flat copper axe still in its superbly preserved haft (44) was among the artefacts belonging to the remarkable 'Iceman' or 'Ötzi', whose 5200-year-old body was discovered high

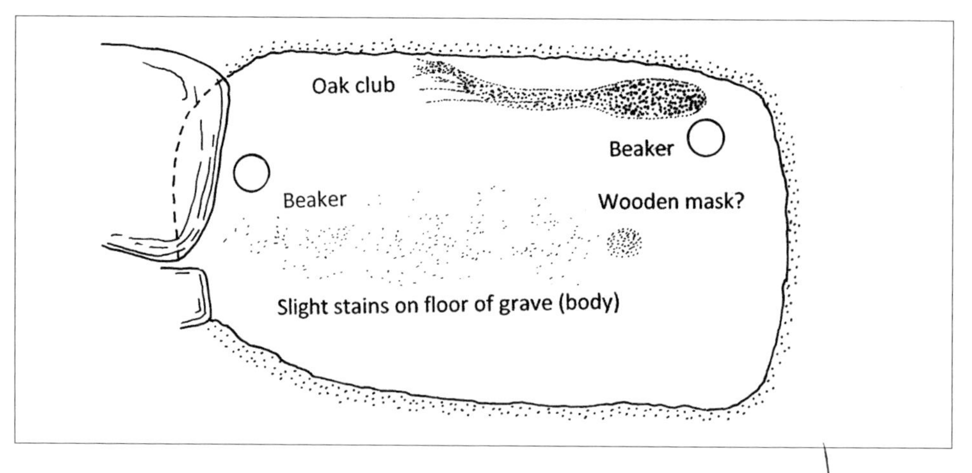

42 Cairnpapple Hill Beaker grave (redrawn after Piggot)

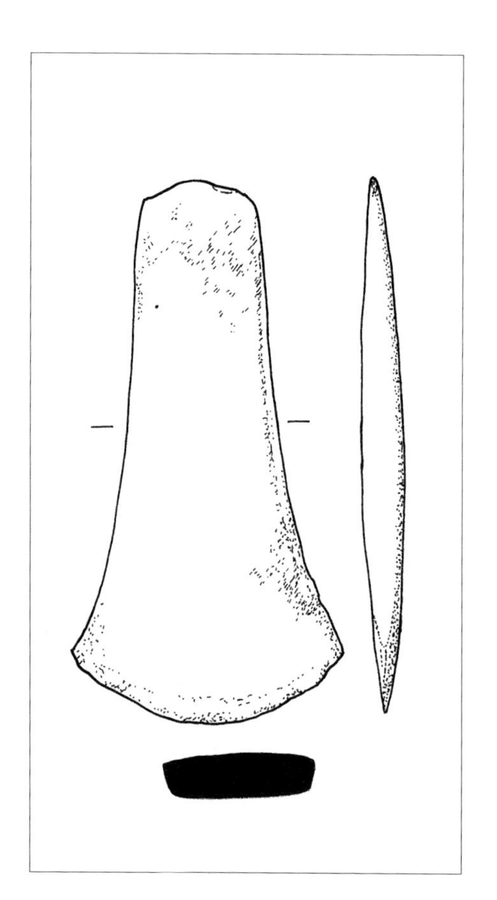

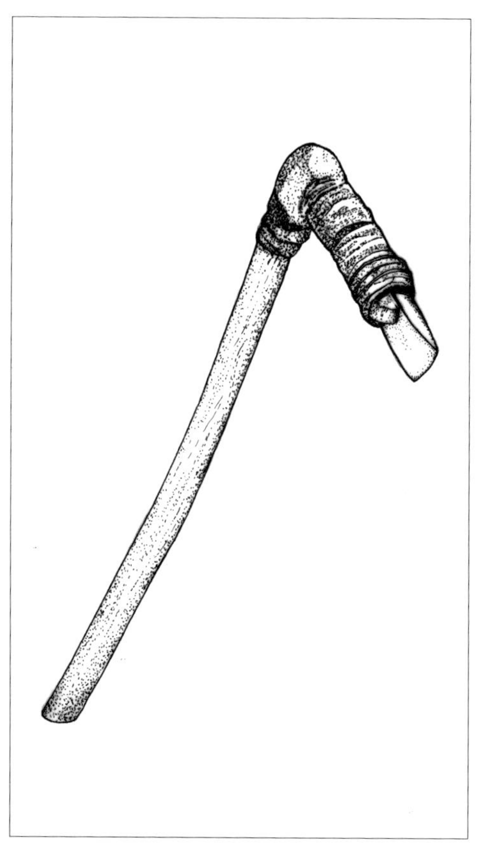

Above Left: 43 Copper axehead from Migdale, Sutherland (redrawn after Britton)

Above Right: 44 Hafted copper axe found with the 'Iceman'

in the Austro-Italian Alps. Interestingly, analysis of his body revealed that Ötzi was involved in some kind of violent struggle and suffered a number of injuries before he was finally killed by an arrow fired into his back (Gostner & Egarter Vigl in Parker Pearson 2005: 19). It may be, as Thomas Loy believes, that Ötzi lost his life because a boundary dispute between different groups had erupted into warfare (Ives 2003).

It is likely that axes signified some level of status and were exchanged in politically motivated transactions by those who had access to what at this time must have been rather exclusive objects (Edmonds 1995: 140). It is also probable that axes were viewed as somewhat magical items by those who lacked the esoteric skills and knowledge of the smiths, who were in effect able to turn rock into beautiful items of shining copper and bronze. However, as Richard Osgood (2000: 78) has rightly pointed out, axes can be interpreted as weapons, though as he also notes (*ibid.*), it is probable that their value lay more in their prestigious nature and their functionality as tools. Nevertheless, this does not preclude them from having at least occasionally been used as combat weapons in the Beaker period, and some of the traumatic injuries observed on skeletal material from Beaker graves might indicate this.

SKELETAL EVIDENCE

We should perhaps turn first to the Amesbury Archer in this brief examination of possible skeletal evidence for warfare in the Beaker period. Intriguingly, analysis of the Archer's skeleton revealed that he had suffered a very serious blow to his left knee. This caused to him to walk with a severe limp and his left leg became wasted as a result of him carrying his weight on his right leg (British Archaeology 2003). Was this injury simply the result of a serious accident (perhaps he had fallen from a horse), or had he been struck in the knee by someone wielding an axe or club?

In addition to the Amesbury Archer, there is the 'Stonehenge Archer' who was discovered by chance in April 1978, when University College, Cardiff, discovered the burial of a young male in the Stonehenge ditch during limited excavations that were being undertaken in order to gather information on the prehistoric environment around Stonehenge (Atkinson and Evans 1978). The young man had a stone bracer next to his lower left arm and there were also three barbed and tanged arrowheads in the grave (45). Somewhat surprisingly, considering their positioning in the grave, like the bracer these were assumed to be the grave goods of a Beaker archer, rather than the weapons that may have killed him (Pitts 2001: 112). However, later analysis of the Stonehenge Archer's skeleton

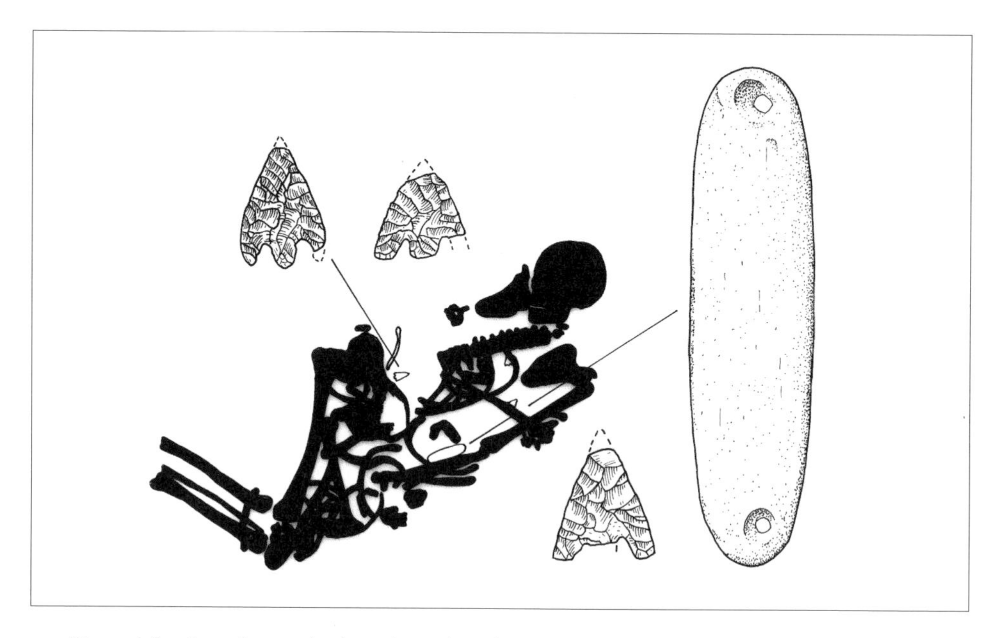

45 Plan of the Stonehenge Archer's burial (redrawn after Edmonds)

revealed that he had been shot at close range. Although all three arrowheads had their tips missing, analysis of wounds on the bones suggested that his assailants may have fired at least four, if not more barbed and tanged arrowheads into his body from the left and right, with the killing arrow probably passing through his heart (*ibid.*: 111-112). The evidence that the Archer had died violently was unequivocal, and although the reason why he lost his life is less clear, the two that seem most likely are that he was either a sacrificial victim, or he had lost his life in combat. Although we cannot discount ritual murder (perhaps of a prisoner of war), it is just as possible that the Stonehenge Archer was given a prestigious burial at this special place because his bravery (and death) in battle were deemed worthy of great honour (Thorpe 2006: 152). It should be noted here that although the Stonehenge Archer is often assumed to belong to the Beaker period, there is a possibility that he actually lived (and died) after this time, as the arrowheads are of the 'Conygar' type, which are often associated with burials of the Early Bronze Age (Gibson 1994: 177).

Although he does not appear to have suffered as fierce an assault as the Stonehenge Archer, an adult male buried in a Beaker flat grave in the centre of a ring ditch at the Neolithic and Bronze Age funerary complex at Barrow Hills in Oxfordshire (Barclay and Halpin 2000: 136-137) was surely killed by the barbed and tanged arrowhead (with an impact fracture at it tip) that was found between his vertebrae and ribs (*colour plate 19*). Another Beaker flat grave found at the site contained the skeleton of adult male, 40-45 years of age, who had a damaged barbed and tanged arrowhead close to his right femur and it is felt that this had been the cause of his death (*ibid.*: 223). Interestingly, this man also had signs of a wound on his right tibia. A cut mark was present on the clavicle of an adult male found in another Beaker grave at the site, although it is perhaps more likely that this mark represents accidental damage during excavation (*ibid.*: 60, 173).

A similar fate to the men at Barrow Hills probably befell the young man whose skeleton was discovered underneath the remains of a round barrow at Fordington Farm in Dorset (dated to c.2350 BC). He had a probable parry fracture to his forearm, but a barbed and tanged arrowhead was also found lying on his hip, and this may well have been fired into his stomach region (Pitts 2001: 286; Bellamy in Thorpe 2006: 152).

Possible hints of armed conflict have been found in two male Beaker burials at Chilbolton in Hampshire (Russel 1990). The younger of the men (aged 20-30 years), who was buried in a mortuary chamber with a rich array of grave goods (which included two pairs of gold basket earrings or hair tresses) had suffered a probable parry fracture to his right forearm, while the older male (probably in his mid 40s) had suffered fractures to the central part of his right rib cage.

Excavations at the well known round barrow at Barnack (Donaldson 1977), which as we have seen, produced a superb bracer decorated with gold caps, also revealed a further 22 burials clustering around the primary Beaker burial that perhaps represented the final resting place of 'an important local chieftain' (*ibid.*: 227). Three of these 'satellite' burials contained individuals displaying skeletal trauma. In Burial 24, a man of 45-60 years had two healed depressed fractures on his skull, one of which bears close similarities to cranial wounds caused by a sling-shot. In Burial 26, a younger male of 30-35 years also had a healed linear wound on his skull, again, probably representing a blow to the head with a weapon of some sort. A possible healed parry fracture was observed on the left forearm of a 30-35 year old male in Burial 28.

Other examples of traumatic injuries from Beaker burial contexts are the adult male from the Pyecombe barrow in Sussex, who had healed fractures to his forearm and collarbone (Butler in Thorpe 2006: 151) and the woman (aged over 50) displaying a likely axe wound to her skull (which was healing at the time of her death), who represented the primary burial in a round barrow containing another three inhumations (Boston in Thorpe 2006: 152). Although she may be 'an unlikely warrior' (Thorpe 2006: 152), we should perhaps be careful not to automatically assume that women did not sometimes go to war in the Beaker period (and indeed in other periods of prehistory), as it is not impossible that some of the other Beaker burials in Britain containing weaponry, actually represent female rather than male burials (Osgood *et. al.* 2000: 3). It has been noted (*ibid.*) that there have been occasions when females actively participated in combat; in the French Revolution and in the Vietnam War for example. From earlier times there is the famous revolt led by Boudica and the women of the Teutonic Ambrones who not only fought against the Roman general Marius, but also attacked the men of their tribe for retreating from battle with the legions.

THE DESTRUCTION OF THE MOUNT PLEASANT AND WEST KENNET ENCLOSURES: EVIDENCE OF WARFARE?

When we consider the question of fortifications in the latter third millennium BC, many scholars are in agreement that that there is little or no evidence to support the notion that they existed during this time. However, at Mount Pleasant in Dorset and West Kennet in Wiltshire, there may be possible evidence not only for fortified sites, but also hints that these sites may have been attacked and destroyed by hostile forces.

Mount Pleasant

Mount Pleasant is situated on a low hill in the Winterbourne Valley near Dorchester and excavations conducted by Geoffrey Wainwright between 1970 and 1971 (Wainwright 1979) revealed that a substantial henge monument had been built here around 2400 BC. As is the case at most other henges, the site had comprised of an outer earthen bank (*c.*4m high) and an encircling inner ditch (*c.*2.5m in depth), with the site occupying an oval area measuring 370m from west to east and 340m from north to south (46). Inside the enclosure in the south-western corner was a smaller monument known as Site IV. This comprised of a circular ditch some 45m in diameter that surrounded four circular arrangements of *c.*170 timber posts, and these were accompanied by settings of standing stones which Wainwright interpreted as a 'Stone Cove' that had replaced the timber structure in the Late Beaker period (1979: 28). The most exciting aspect of the site however was the discovery of a narrow but deep palisade trench (*c.*1m wide and 3m in depth) running just around the inner edge of the main ditch. Examination of the ditch revealed that it had once contained a massive wooden palisade (constructed *c.*2100 BC) consisting of some 1600 oak timbers that averaged 40cm in diameter and reached at least 6m in height (the full height of the posts had therefore been an impressive 9m, or more). There were also two very narrow entrances in the east and north of the palisade and the one to the east had been flanked by massive timbers measuring 1.5m in diameter (47).

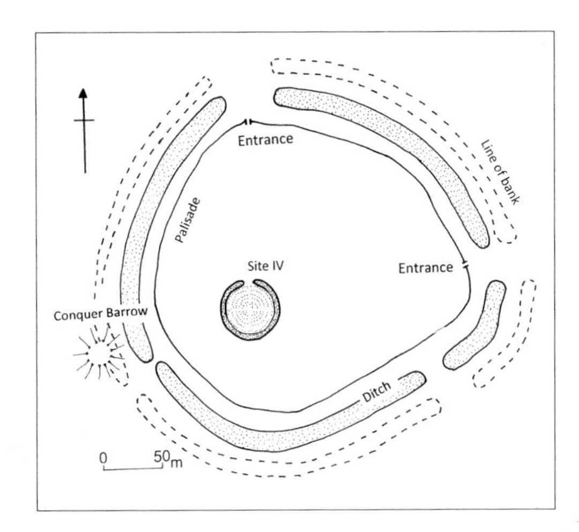

46 Plan of Mount Pleasant (redrawn after Pitts)

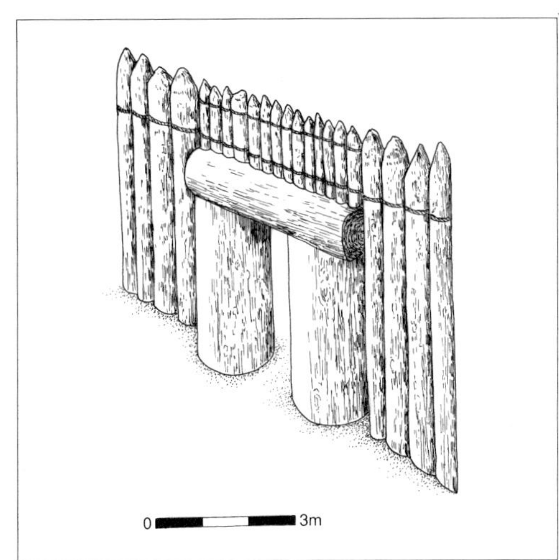

47 Reconstruction of massive timber gateway in Mount Pleasant palisade (redrawn after Wainwright)

It was clear from the evidence discovered in the trench that the palisade had been deliberately set on fire and destroyed; burnt posts were revealed as preserved lumps of charcoal and at some points of the palisade circuit, the fire had been so fierce that the posts had burnt all the way down to leave cores of ash and burnt chalk. In other places, the posts had been dug out and removed, presumably to be reused elsewhere. In addition to the evidence for the fierce fire that had destroyed the palisade, there was also a large amount of stone debris and charcoal found in the ditch of the timber circle. Although it is not conclusive, this evidence may represent the destruction of the standing stones at the same time as the massive timber palisade was deliberately set on fire and slighted, rather than waste from their shaping when they replaced the decayed posts of Site IV as was originally suggested by Wainwright (Pitts 2001: 255).

How can this evidence be interpreted? Geoff Wainwright (1979: 241) argues that it is difficult to envisage the huge palisade (which he felt surrounded a settlement) as anything other than defensive, though he admits that we cannot say for sure whether its destruction was a voluntary or aggressive act. Likewise, Mike Pitts (2001: 68) feels that the palisade was clearly a military structure that took defensive advantage of the contours of the hill on which it was situated, and furthermore, that its destruction only serves to emphasise this. In other words, the site appears to have been attacked and destroyed by an enemy force. On the other hand, Thorpe (2006: 151) argues that it is unlikely that the relatively low-lying hill on which the enclosure is situated would be suitable for a defended settlement and argues that there are much more suitable locations for the building of one in the vicinity, such as Maiden Castle. In a similar vein, Mercer (2006: 130) feels that as the site sits askew the hilltop it could not have been defensive in nature and points out that there are neither victims nor weapons of war found at the site. Thus he concludes (*ibid.*: 131) that the palisade may well have been deliberately set on fire by those responsible for its construction – perhaps in some type of religious ritual or ceremony. Nevertheless, while there may be no evidence of war casualties and the weapons that killed them, this could be because the Mount Pleasant enclosure was a place built for the dead rather than the living, and there may have been no living people here when it was deliberately burnt down and destroyed by an enemy force (Pitts 2001: 293).

The West Kennet enclosures

These two adjacent palisade enclosures lie in the West Kennet Valley, an area long famed for its superb prehistoric monuments such as Silbury Hill and the West Kennet Long Barrow. Both enclosures, which may have been contemporary, featured huge timber posts measuring some 8m in height, with post diameters measuring 30-50cm and there were other associated features such as radial palisade and fence lines (48). As to gaining a firm date for the enclosures, it has to be said that we are largely groping in the dark, because as Mike Pitts has pointed out, the dates that have been obtained from them are so mixed up that they are practically useless in this regard (pers. comm.). Nevertheless, Mercer (2006: 131) notes a date between 2600-2100 BC for their construction and therefore, the likelihood is that they were built during the Beaker period. It was also clear that both enclosures had been destroyed by fire, as the post-holes or 'post-pipes' which marked the existence of the timbers of the original palisades were rich in charcoal. It seems highly unlikely that the conflagrations that destroyed them were accidental. Therefore, we could again be looking at the deliberate attack

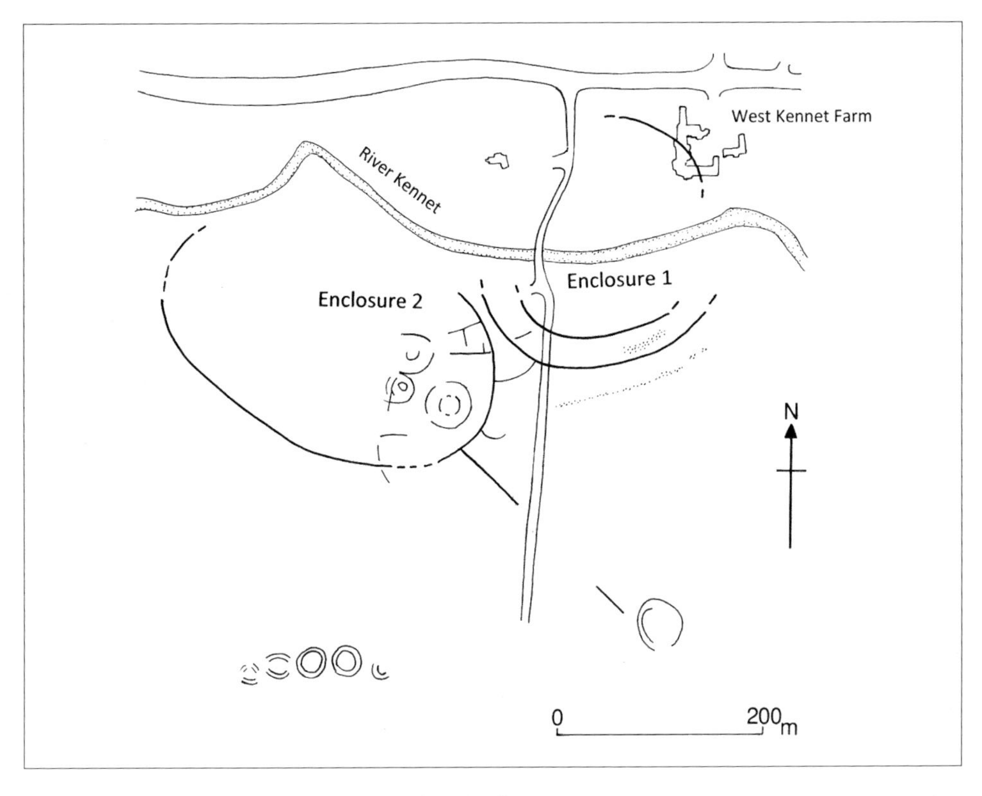

48 Plan of West Kennet enclosures (redrawn after Pitts)

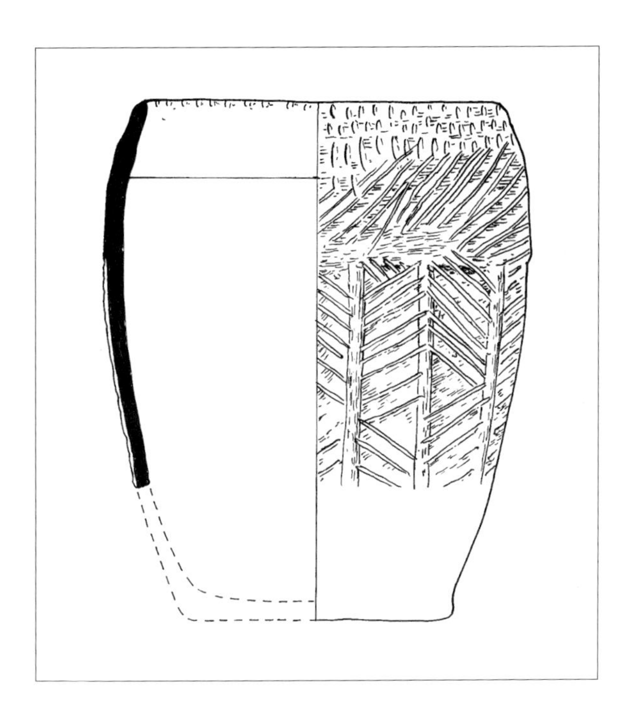

49 Grooved Ware vessel (redrawn after Pryor)

and destruction of fortified sites, although the huge size of the enclosures and their low-lying position in the Kennet Valley perhaps argues against them having had a defensive purpose (Whittle 1991b: 261).

However, whilst the Mount Pleasant and West Kennet palisade enclosures may not have been deliberately designed with defence in mind (although the evidence at the former site is perhaps more suggestive in this regard), this does not preclude them from having been deliberately destroyed in acts of aggression. It could be tentatively suggested that this aggression relates to tensions between Beaker groups and users of the equally distinctive 'Grooved Ware' pottery (49). The latter appear to have belonged to a widespread Late Neolithic cultural tradition (appearing *c.*3000 BC) that is closely linked to the building and use of large ceremonial monuments (in particular henges), and which continued to co-exist alongside the Beaker culture until it finally faded away in the last centuries of the third millennium BC. Therefore, in the latter half of the third millennium BC, we have a period 'of significant cultural overlap' (Needham 2008: 19), with arcane and conservative Grooved Ware practices continuing to take place during a time of great innovation and change that was ushered in with the arrival of the highly distinctive Beaker 'package' with its metalworking and emphasis on individual prestige and weaponry. Thus, the marked contrasts that are readily apparent in the archaeological record of the latter half of the third millennium BC indicate that the Grooved Ware and Beaker cultural complexes appear to represent the contemporary existence of two radically different political and ideological groups (Edmonds 1995: 139). The fact that Grooved Ware and Beaker pottery are often found on the same sites (Needham 2008: 19), undoubtedly reveals that there were many peaceful interactions between Grooved Ware and Beaker communities. However, in time, relations between the two cultures may have become increasingly strained as the Beaker culture began to assert a dominant position in society, perhaps because of the continuing arrival of small numbers of Beaker immigrants and because more people were marrying into Beaker communities with their new and attractive ways of life (*ibid.*: 22). Subsequently, localised friction and squabbles may have become more common as the elite of the Grooved Ware society watched their once secure position at the head of Late Neolithic society being slowly eroded by the Beaker culture (*ibid.*) and such tensions could have flared up into warfare between the two groups.

If the evidence does indeed point to acts of armed aggression at Mount Pleasant and the West Kennet enclosures, do we have any signs as to which group was actually behind them? In regard to the latter site, the short answer is probably no. However, the fact that the largest collection of Grooved Ware from the district along with petit tranchet arrowheads (which are also closely linked to the Grooved Ware complex) were found at the enclosures (Pitts 2001: 283) surely reveals who was behind their construction. It could therefore be suggested that if the two enclosures were indeed deliberately raised to the ground by an enemy force, this act would probably not have been carried out by people of the same 'culture' and that those responsible may have been members of a Beaker group. However, it could still be the case that it was a rival Grooved Ware group who burnt down the enclosures as it seems unlikely that relations between different Grooved Ware communities were always peaceful (in this respect, we should perhaps recall the femur displaying arrowhead impact injuries from Durrington Walls, mentioned in Chapter Two).

Turning to Mount Pleasant, it is unsurprising that sherds of Grooved Ware are found in the earliest levels of its ditches and it is very likely that people of the Grooved Ware culture were responsible for the construction of this henge, and probably its interior timber and stone monument as well. Intriguingly however, Beaker sherds are found in the latest levels at the site, and perhaps significantly, are found mixed in with the stone debris found in the ditch of Site IV (*ibid.*: 255). Could it be then, that these simple sherds of pottery bear witness to the destruction of a sacred Grooved Ware site, which was erected in an act of what has been called 'monumental aggrandising' (Mike Pitts pers. comm.), but was subsequently destroyed by those who followed the Beaker way of life? Or was it vice-versa, with Beaker users erecting the massive palisade and replacing the timber ring of Site IV with a circle of standing stones (Parker Pearson 1993: 88), which were subsequently destroyed in an act of aggression by 'a disaffected rump' of Grooved Ware society (Needham 2008: 22), who were not prepared to watch their traditional way of life (and power) being replaced by the innovative Beaker culture? Such conundrums reveal the complexities that face those trying to make sense of the prehistoric past.

THE EARLY BRONZE AGE C.2000-1500 BC

As we move from the Beaker period into the early centuries of the second millennium BC, it is evident that the weapons introduced with the arrival of the Beaker phenomenon continued in use into the Early Bronze Age, although battleaxes were now more common and different types appeared alongside new styles of dagger (*50* & *51*). Bronze axes were also now given additional flanges to aid hafting and their blades were expanded (*52*). During this period the bow and arrow was still in use and fine barbed and tanged arrowheads of the 'Breton' type also appear in graves. Some of these are so thin in section and so beautifully made, that they seem to have been purely ceremonial or symbolic weapons to be placed with the deceased. While this appears likely in some cases, it has been

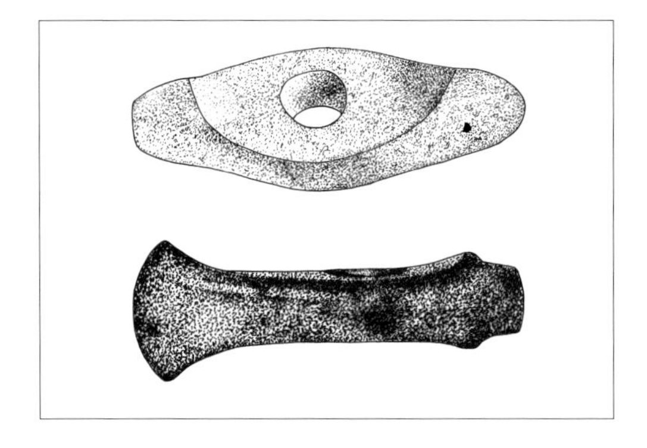

50 Early Bronze Age battleaxes

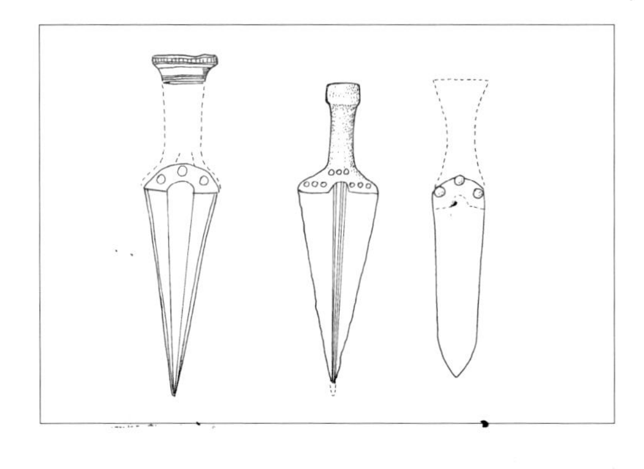

51 Early Bronze Age daggers (redrawn after Piggot)

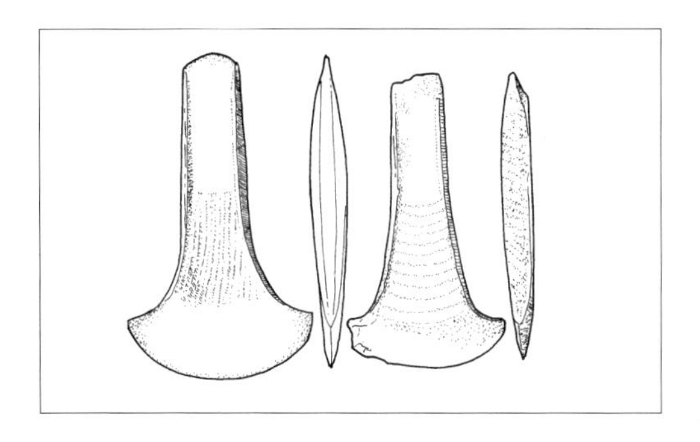

52 Early Bronze Age axes

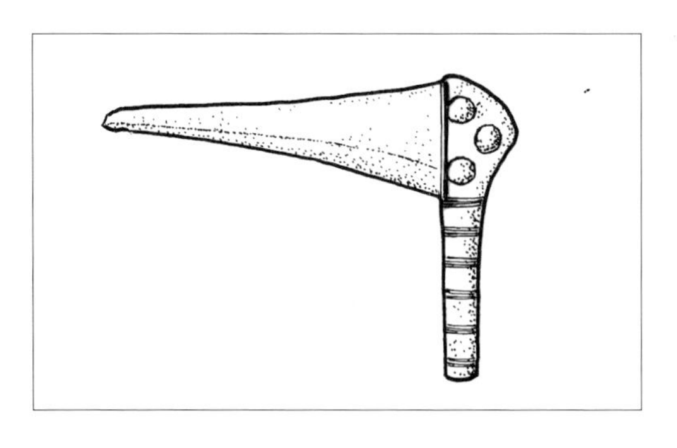

53 Halberd from central Europe

argued (Jiménez & Sánchez Romero 2006: 188) that similarly made arrowheads found in Copper Age Iberia, would have actually travelled greater distances and that their increased sharpness made them more deadly as penetrating weapons which would break easily inside a human body.

We must also consider the halberd among the weapons that were used during the Earlier Bronze Age. These rather puzzling implements probably emerged in Ireland around the end of the third millennium BC (Mercer 2006: 125) and basically consist of a large dagger-like blade that would have been fixed at right angles to a long wooden shaft (53). Although their design made halberds vulnerable to breakage and rather unwieldy if used in combat, use damage commonly observed on their hafting plates indicates that they were probably used in a similar fashion to medieval poleaxes, with their vicious blades used to deliver the coup de grace after an initial blow with the wooden shaft (O'Flaherty *et. al.* in Thorpe 2006: 153). Depictions of what appear to be warriors wielding halberds can be seen among the many hundreds of carvings found on the famous Naquane rock at Valcamonica in northern Itlay (Toms 2000: 111). At the end of the Early Bronze Age, metal spearheads make their first appearance (Osgood 2000: 25; Harding 2000: 281) and it is clear that these became a major weapon in the succeeding centuries of the Later Bronze Age. As we will see in the next chapter, the site of Tormarton provides us with graphic evidence of the terrible effectiveness of the Bronze Age spear as a killing weapon.

WARLORDS AND WARRIORS IN THE EARLY BRONZE AGE OF WESSEX

Any discussion of warfare in the Early Bronze Age must also take into account the barrow burials of the famous 'Wessex Culture' (Piggot 1938) which dates to the first centuries of the second millennium BC. Although it now appears unlikely that these burials actually represent the existence of an actual culture *per se*, it is obvious that they mark an elite section of society which had wide ranging contacts (particularly with Britanny). It is probable that in some cases, they represent the final resting places of powerful chieftains or warlords and their warrior retinue who lived 'in a heroic age where authority was obtained through warfare and displayed through the display of exotic and unique regalia' (Burl 1989: 154). The most notable burial of the Wessex Culture is that found underneath the famous 'Bush Barrow' on Normanton Down, some quarter of a mile from Stonehenge. The barrow was 'excavated' ('dug into' would be a better term) by William Cunnington in the summer of 1808 who discovered the skeleton of a tall and robust male lying on his back, accompanied by some of the finest prehistoric artefacts ever found in Europe. On his chest, was a superbly made sheet gold lozenge that had been beautifully decorated and which had presumably been attached to the man's tunic when he had been buried, as there were two perforations at either end. In addition to this larger lozenge, a much smaller version was found close to the man's right hand and a sheet gold plaque which has usually been interpreted as a 'belt-hook' was also discovered. As Aubrey Burl (1981: 158) has said, in his usual inimitable style, 'this was not the finery of a popinjay'; also included in the grave was a flanged metal axe, a fine, egg-shaped polished macehead made from a rare limestone found in Devon and two very large daggers featuring traces of their wood and leather scabbards (54). One was made from copper, and one from bronze, with the larger weapon reaching a fearsome 33cm in length. Remarkably, close analysis of the copper dagger revealed that its handle had been decorated with thousands of tiny strands of gold wire, which formed a zigzag pattern in the handle. Lying behind the man's head were several rivets, which have been interpreted as both the remains of a

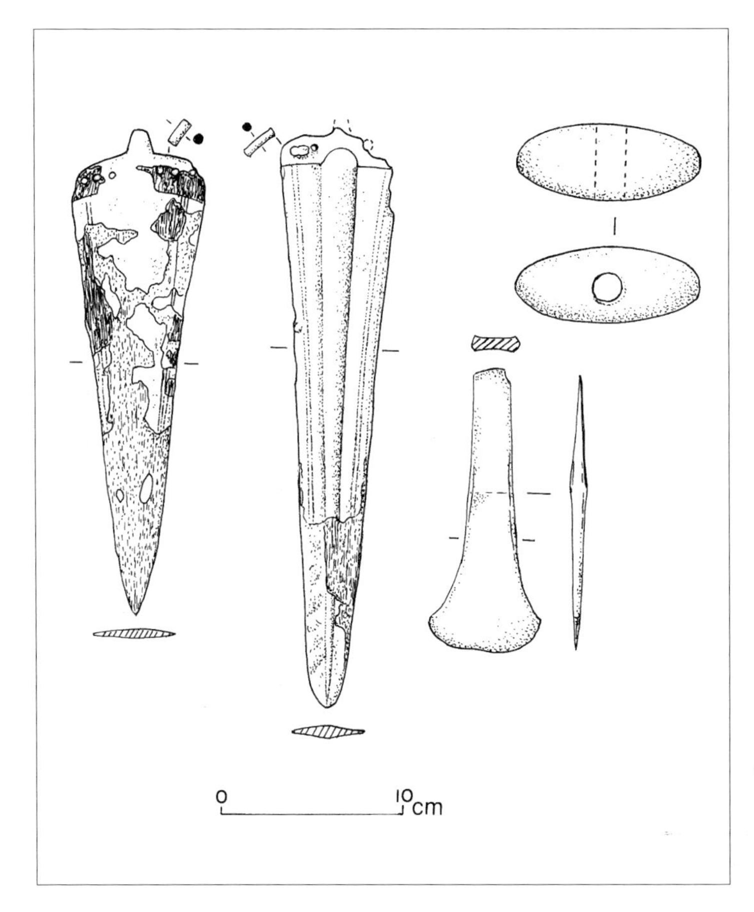

54 Daggers, macehead and bronze axe from Bush Barrow (redrawn after Annable & Simpson)

shield and a surveying instrument for the laying out of Stonehenge. However, it has been argued (Megaw & Simpson 1979: 212; Burl 1981: 158) that the small size of the rivets would have been too small for such purposes and therefore it is more likely that they represent the remains of a studded leather helmet. More recently, though, both the shield and helmet theories have been labelled 'unconvincing' (Needham *et. al.* 2008b: 14) and it has been proposed that the rivets actually represent the remains of another dagger with studded hilt (*ibid.*).

Whilst it is unlikely that the Bush Barrow shield ever existed, an intriguing find from Kilmahamogue in County Antrim reveals that in Ireland at least, shields

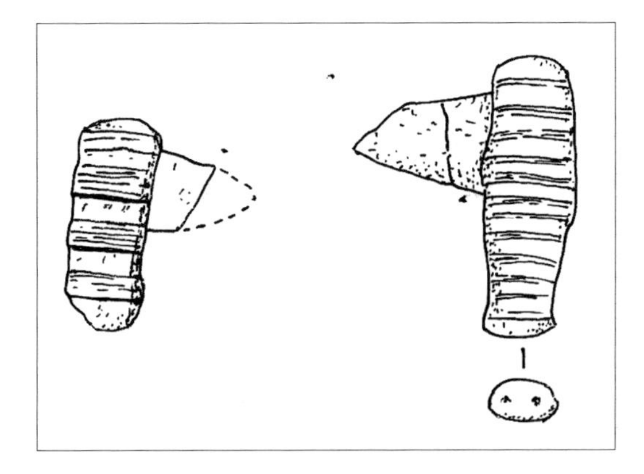

55 Miniature halberd pendants (redrawn after Piggot)

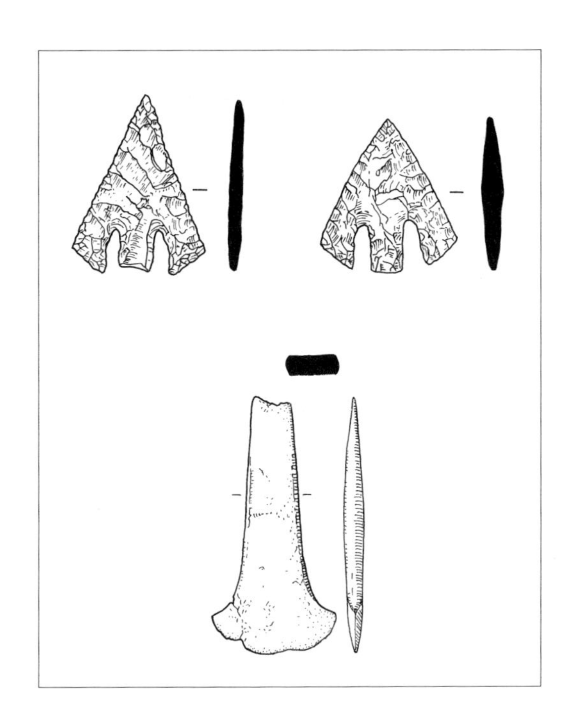

56 Arrowheads and bronze axe found at Breach Farm, Glamorgan (redrawn after Grimes)

were in use in the Early Bronze Age. This find took the form of a wooden shield 'former' or mold that had been used to produce V-notched leather shields and radiocarbon dates obtained on it gave a rather surprising date of *c.*1950-1540 BC (Hedges *et. al.* 1991: 129). It is therefore perhaps possible that in Ireland at any rate, swords were present at an earlier date than commonly assumed, though of course shields may have been used as a defence against other weapons.

As well as signalling his high status, the strong militaristic overtones of many of the items that were buried with the man under Bush Barrow imply that warfare is likely to have played a significant part in his world. It is possible that cattle loomed large behind this warfare, as a number of scholars (e.g. Ashbee 1960; Fleming 1971) have proposed that the people of the Wessex Culture were pastoralists and have pointed out that pastoralism in non-state societies stimulates the rise of warrior societies and warfare. Therfore, raiding (for both cattle and perhaps also people) may have been common and these raids, together with possible disputes over 'unlawful' incursions into grazing territories, may have sparked more significant episodes of warfare.

In addition to the richly furnished Wessex graves containing gold and weaponry, there are contemporary barrow burials from Wessex which are quite conceivably monuments to warriors who were a rung below the elite on the social scale. For example, at the Amesbury 15 bell-barrow in Wiltshire, two bronze daggers accompanied the burial, whilst the well known 'King Barrow' near Stonehenge may have held not only a bronze dagger but also – judging from an eighteenth-century account of Lord Pembroke – what appears to have been a large bronze halberd (Burl 1989: 165-166). Miniature halberd pendants have also been found in Wessex Graves, such as those found in richly furnished burials discovered at Manton and Normanton in Wilshire (55). In addition, to the 'dagger-graves', there are also several Wessex barrows containing battle axes and examples include Wilsford G8 and a barrow found on Windmill Hill (the site of a famous Neolithic causewayed enclosure) in Wiltshire (Megaw & Simpson 1979: 215). We should also remember that in many other areas of Britain during the Early Bronze Age, there are burials containing fine artefacts and weaponry, which again hint at the existence of stratified warrior societies. A notable example is provided by the Breach Farm Barrow, Glamorgan, where a probable dagger, a bronze axe, two sandstone arrowshaft smoothers and a collection of superbly made Breton-type arrowheads were found among the grave goods (56). Most importantly, however, we should not lose sight of the fact that to their enemies, the warriors of Early Bronze Age Britain would not have been seen as 'heroes' but rather, would have been feared as agents of violent death, destruction and grief.

SKELETAL EVIDENCE

This brief examination of the possible evidence for armed conflict in the Early Bronze Age begins with two burials that bear close comparison with that of the Stonehenge Archer. The first comes from the fascinating timber circle excavated at the Sarn-y-Bryn-Caled ritual and ceremonial complex near Welshpool, Powys (Gibson 1992, 1994). In the centre of the timber circle, two successive cremation burials were discovered in a rectangular pit, and among the bones of the earlier cremation burial at the base of the pit were four superbly made barbed and tanged arrowheads. Interestingly, two were missing their tips and had impact fractures, and in addition to this, although the arrowheads were turned white by the intense heat of the funeral pyre, they were not shattered as would be expected if they were grave goods. Therefore, as Alex Gibson has plausibly suggested, the evidence implies that the arrowheads were the cause of death and were actually in the body when it was laid on the funeral pyre. However, as with the Stonehenge Archer, there is some uncertainty to the actual date of the burial, as the radiocarbon date of *c.*2400-2300 BC – suggesting that the individual lost his/her life in the Beaker period – does not tally with the arrowheads which were again of the Conygar type. Although the evidence is not as conclusive as at Stonehenge, it is still probable that the individual at the Sarn-y-Bryn Caled timber circle was killed by the arrowheads which were found accompanying the burial. Again though, we are faced with the question of whether this individual was a sacrificial victim, or someone who was accorded the great honour of being buried in an important religious monument because of their bravery in warfare.

Another such conundrum faces us with the probable Early Bronze Age burial found at a cremation cemetery near Grandtully in the Tay Valley, Scotland (Simpson & Coles 1990). Five Conygar arrowheads were found with the cremation of a young adult (possibly female) in a pit and again, Gibson (1994: 87) has credibly suggested that they were actually in the body prior to its burning.

A much earlier find from an undated burial cairn known as 'Twr Gwyn Mawr' at Carno in Montgomeryshire (Davies 1857) also points in a similar direction as the above burials. Here, two barbed and tanged arrowheads were found among a cremation burial that lay underneath the Cairn. As Thorpe (2006: 152) notes, the published illustration of the arrowhead, appears to show that one has an impact fracture as its tip, although unfortunately the arrowheads have been lost, so this cannot be confirmed.

A similar discovery was made at Ballymacaldrack, County Antrim, where a rough barbed and tanged arrowhead with broken tip was found in association with a probable female cremation burial in an Early Bronze Age Collared Urn (Tomb & Davies 1938). Professor Walmsley who examined the many fragments of bone found in the Collared Urn felt 'that it may have been the cause of death, as the bones have all the signs of vigorous and healthy development' (*ibid.*: 221).

At an Early Bronze Age bell-barrow near Sutton Veny in Wiltshire a secondary satellite burial found on the edge of the barrow (Johnston 1980) provided intriguing but disputed evidence for warfare. The skeleton found in the grave probably belonged to a man aged around 25 years old and in addition to an earlier healed fracture from a depressed wound, there was 'a particularly violent head-wound' (*ibid.*: 38) that had evidently caused the young man's death. It was felt that the killing blow had been made by a sword and so the burial was dated to the Late Bronze Age when swords made their first appearance in Britain. However, the injuries may result from a blow from a heavy, blunt object and it is even possible that the burial dates to the Anglo-Saxon period (Thorpe 2006: 154).

Underneath one of the round barrows found in a barrow cemetery near Cheltenham, what appears to have been a multiple burial was found, and one of the skulls displayed fractures which could have been caused by a sling stone (Bird 1865: 69), although as has been rightly pointed out (Schulting & Wysocki 2005: 127), we must treat this evidence with a high degree of caution.

A skeleton found underneath a bowl barrow at Amesbury might represent a victim of war, or alternatively a ritual mutilation; its right arm was missing, the other arm was separated from the body and the hand had been severed, whilst the ribs were crushed, and the skull lay some distance from the body (Ashbee 1960: 79). A skeleton found in a disc barrow not far from this intriguing find, showed signs of similar treatment (*ibid.*)

Leslie Grinsell (1941: 97) also mentions in a survey of round barrows in Wessex that one of them featured an inhumation burial in which an individual's head had been severed, though of course this does not necessarily indicate warfare.

Moving far North from Wiltshire to the headland of Cnip on the lovely Isle of Lewis in the Western Isles of Scotland, a man aged around 40 years was discovered in a stone-lined grave or cist with an undecorated pot (Close-Brooks 1995; Dunwell *et. al.* 1995; Cowie & Macleod 2002). A sample of bone was radiocarbon dated and revealed that the man had been buried around 1700 BC. Analysis of his skeleton showed that he had healed but severe and extensive injuries to the right side of his face which probably reveal that the man had been clubbed in the right side of his head. It is possible that he received these injuries in an episode of warfare that was related to competing claims over good quality farmland that was becoming scarcer at this time (Cowie & Macleod 2002: 35).

CHAPTER 4

THE LATER BRONZE AGE C.1500-750 BC

There can be little doubt that warfare was a significant feature of life in prehistoric Britain during the Later Bronze Age (comprising the Middle and Late Bronze Age) as there is plentiful evidence from the archaeological record which is clearly militaristic in nature. In addition to a proliferation in weaponry which has been said to mark the 'first arms race' (Osgood 2000: 23), we also see the appearance of hillforts in many parts of the British landscape during this period, although of course, as we have seen, 'hillforts' were not unknown in the Neolithic. This chapter will begin with a look at the fascinating evidence found at the remarkable site of Tormarton, which not only yielded likely casualties of war from the Later Bronze Age, but also forcibly brought home the truth that life in prehistoric Britain could be very brutal.

TORMARTON: EVIDENCE OF TERRITORIAL WARFARE?

The site of Tormarton lies on West Littleton Down in the south-western Cotswolds and the grim secret that it had hidden for some 4000 years was discovered in July 1968, during the digging of a gas pipeline across the Down (Knight *et. al.* 1972). The remains of at least three young males were found in the disturbed soil and subsequent analysis of the bones showed that 'An act of extraordinary violence' (Osgood 2000: 21) had indeed taken place at Tormarton and that two of the men had suffered terrible weapons injuries. The oldest male (*c.*35 years) had been stabbed from behind at least once with a spear and this had been thrust through his body into his pelvis, where it left a clear lozenge shaped hole (57). The younger of the two men (*colour plate 20*) who was in his 20s, 'had suffered wounds that are shocking to anyone that sees them' (Osgood 2006: 331). Like his older counterpart, he had also been speared in the pelvis, with the bronze spearhead thrust into the body and twisted so that it snapped from its shaft and

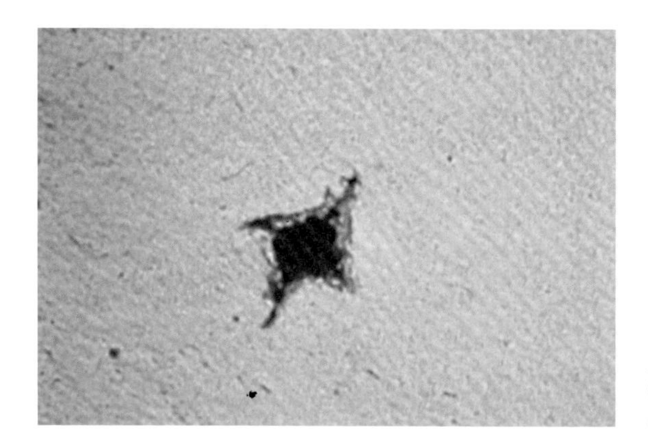

57 Lozenge-shaped hole in older male's pelvis from Tormarton (Richard Osgood)

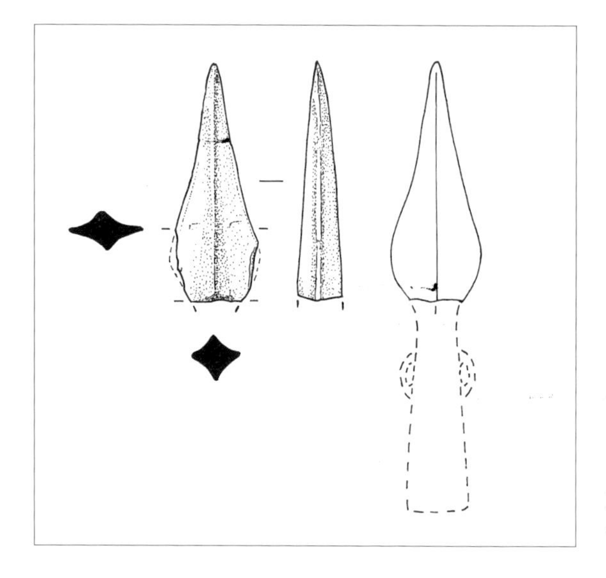

58 Spearhead found in lumbar vertebrae of younger male at Tormarton (redrawn after Osgood)

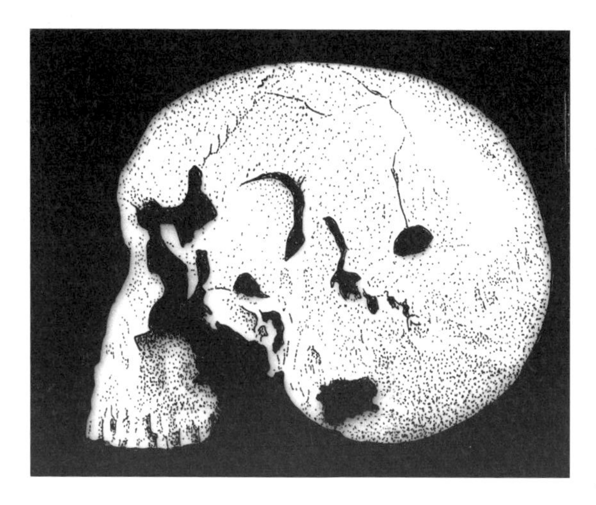

59 Skull of younger male from Tormarton, displaying probable spear-hole

remained in the bone (*colour plate 21*). Another spear had been thrust into his back with such force, that it had pierced his lumbar vertebrae and severed his spinal cord, which would have paralysed him instantly. Again, the spearhead had broken off and remained embedded in his skeleton (*colour plate 22*). Examination of the spearhead suggested that it was of a type used in the Middle Bronze Age (*58* & *colour plate 23*) and a radiocarbon date of *c.*1315-1045 BC taken from the person whom it had killed indicated that the terrible violence at Tormarton had occurred at the end of this period. In addition to being stabbed at least twice by spears, the younger man was also probably stabbed in the head with a spear, as evidenced by the small circular hole seen on his skull (*59*). It is quite probable that this blow finished him off as he lay stricken on the ground, paralysed and in agony.

During 1999 and 2000, Richard Osgood and a small team returned to Tormarton to conduct further investigations at the site (Osgood 2005; 2006). They were not disappointed in their efforts, as they found a large, linear V-shaped ditch (in the same place as the human remains found in 1968) that was around 3m wide, 1.5m deep and some 70m in length (*60*). The ditch had been deliberately filled with limestone slabs and it was evident that this had been done quickly in a single event. Large quantities of human bone were also found in one of the trenches, though this material was not as well not preserved as that found in the original excavation. Nevertheless, analysis of this bone suggested that it had probably come from a further two individuals and thus the likelihood is that five males (who ranged in age from around 11 to 38 years) are represented at Tormarton. Whilst obvious traces of violent trauma were only recorded on

60 Linear ditch at Tormarton (Richard Osgood)

two individuals at Tormarton, it is very unlikely that the others found in the ditch died a peaceful death and this site probably bears witness to a savage act of violence in which a group of five young males (which included at least one child) were killed and dumped into the ditch that was then filled in by those who brutally killed them.

It appears that the human remains found at Tormarton bear witness to the violent death of five individuals, but can we come any closer to the 'human story' which lay behind this terrible event that took place in a quiet corner of the Gloucestershire countryside some three and a half thousand years ago? Although we can of course never be absolutely sure of our interpretations when dealing with the mute evidence of prehistory, it appears likely that at Tormarton, 'the linear ditch is central to events or, at least, the culminating event' (*ibid.*: 143). During the Later Bronze Age, the landscape became increasingly divided up into separate areas of farmland and settlement. These areas are often delineated by linear boundaries and imply the existence of separate groups marking out their territories, perhaps because of increasing competition over resources and agricultural land (*ibid.*).

The most notable example of this Later Bronze Age demarcation of territories is provided by the famous Dartmoor 'reaves', which primarily date to the Middle Bronze Age, although their origins probably lie in the Early Bronze Age. These low stone walls (some of which were originally probably quite substantial) are still visible on many parts of Dartmoor today, and in some cases they stretch for many miles across the landscape. It is hard to resist the idea (Fleming 1978: 97) that they reveal the existence of several different 'socio-political units' who although living in separate territories, probably used the upland areas of Dartmoor as common grazing land. Associated with the reaves, there are major enclosures (e.g. Grimspound, Cholwich Town and Ryder's Rings) with massive stone walls that are built in commanding positions in the valleys in which they are located, and these could well mark the fortified settlements of the separate groups in this region (*ibid.*: 109-110). Whether actual warfare broke out between these groups or between natives and incomers is unknown, but hints of it have been found at places such as Fice's Well where a well-preserved rapier was found, and Bloody Pool, which produced a hoard of fine spearheads (Pettit 1974: 162-163). Also of note are the hundreds of kilometres of linear ditches that can be found in many parts of the Wessex region, some of which are likely to represent territorial boundaries of different Late Bronze Age groups (Cunliffe 2004). Again, at times, these territories may have been violently disputed by different communities.

Although at Tormarton the boundary seems to have been more 'symbolic' than real (Osgood 2006: 337), it is probable that its digging represents the attempts of a group to lay claim to an area of land suitable for farming (*ibid*.: 339). As we have seen, it was an attempt that appears to have ended in terrible failure, with the brutal deaths of members of a Bronze Age community. It is not impossible that other members of their community were killed as result of this probable dispute over land, even though they have not been recovered from the archaeological record.

WEAPONS FOR 'SHOW' AND WEAPONS FOR WAR

Having just considered the evidence from Tormarton, it seems appropriate to begin this brief discussion of the weaponry of the Later Bronze Age with spears. Whilst spears were doubtless used for hunting wild animals in the Later Bronze Age, evidence such as that found at Tormarton proves that they were deadly weapons in terms of taking human life and it is more probable that they were designed with this express purpose in mind.

Although a great variety of different shapes and sizes of spearheads can be seen from the Later Bronze Age in Britain (*61*), they can basically be divided into larger and smaller groups with the former assumed to be 'lances' for thrusting or stabbing at close quarters and the latter, 'javelins' to be hurled at opponents from a longer range (Osgood 2000: 25; Harding 2000; 281). However, as has been stressed (Osgood 2006: 338-339), the evidence from Tormarton shows that it is probably unwise to draw such rigid distinctions when considering the spearheads of the Later Bronze Age; how they were used may have depended as much on the nature of the encounter in which warriors found themselves

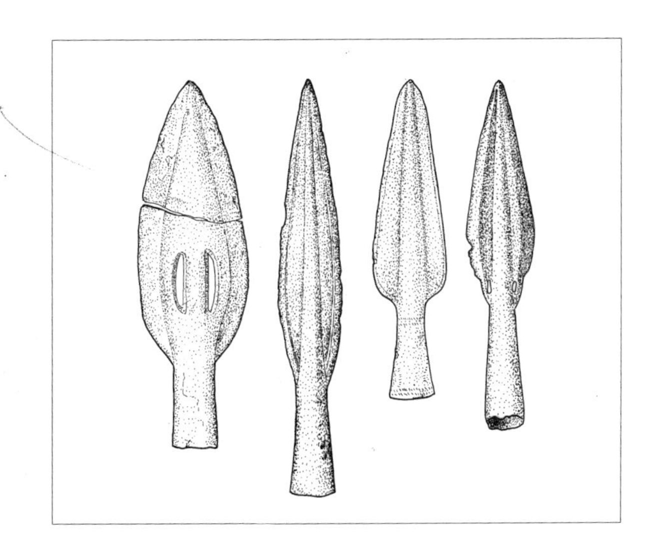

61 Later Bronze Age spears (redrawn after Burgess *et. al.*)

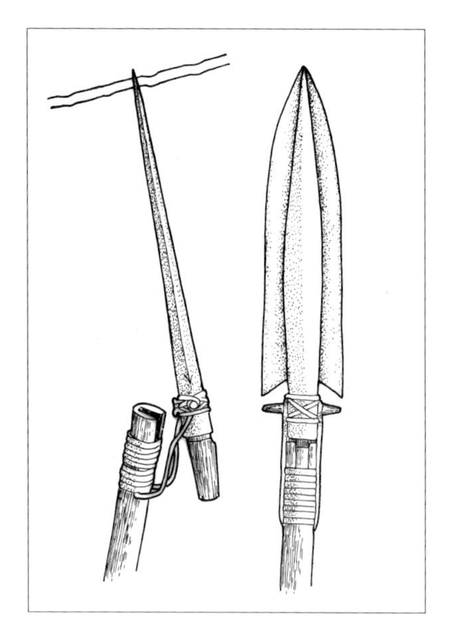

62 Spearhead from North Ferriby (redrawn after Bartlett & Hawkes)

embroiled. Nonetheless, while this argument undoubtedly rings true, the large number of smaller javelin-type spearheads found throughout Europe suggests that in some cases at least, the first stage of combat consisted of the throwing of spears at close range (perhaps around 30m), with the more deadly fighting occurring at close quarters (Harding 2000: 283). It may be that those who had thrown their spears took cover behind an advancing line of spearmen, only emerging to engage with the enemy as they clashed at close-quarters (Harding 1999: 165-166). Whatever the reality is in regard to the tactics employed by spearmen in combat, the fact that many different types of spears were used in the Later Bronze Age lends credence to the idea 'that spearmen became an important fighting force', particularly in the Late Bronze Age, *c.*1100-800 BC (Lynch 2000: 108). This idea is lent further support by evidence gleaned from Jill York's (2002) study of Bronze Age metalwork deposited in the Thames and its tributaries, which suggested that 82 per cent of the spearheads had actually been used, with many showing broken or stubbed tips and chipping of their edges. It would be unrealistic to argue that all of this damage was incurred in hunting or as a result of post-depositional movement of the spears in rivers.

A spear found at North Ferriby in Yorkshire should also be mentioned here. It has been suggested (Bartlett and Hawkes 1965: 373) on the basis of its design that it was used in a similar manner to a Roman *pilum* and worked on the principle of encumbrance, with the heavy spearhead made to detach from its shaft and lodge in an enemy's shield (62). Although as Osgood (1998: 15) notes, we need more examples of such spears before we can say for sure that they

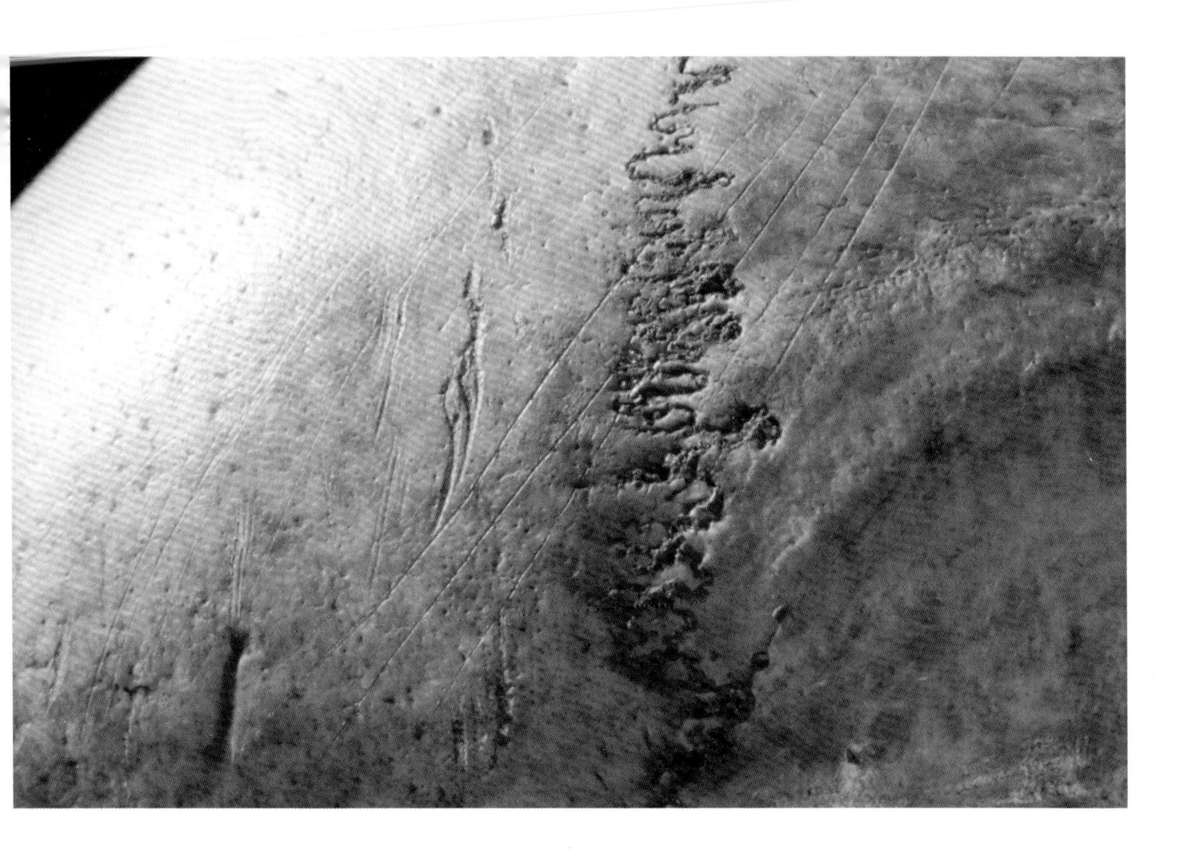

Above: 1 Close up of adult skull cap from Gough's Cave showing cut-marks (Natural History Museum)

Right: 2 Skull of Cheddar Man – note the large abscess hole (Natural History Museum)

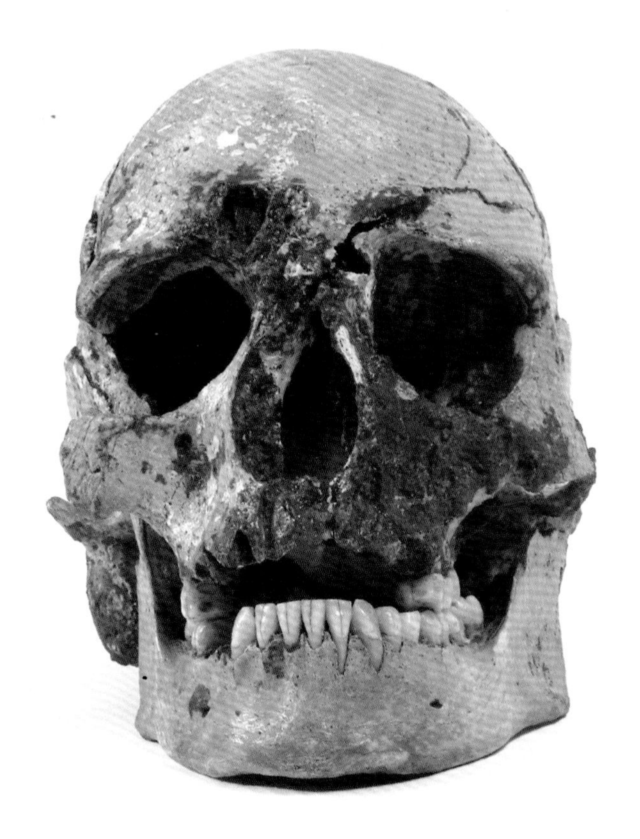

3 Neolithic axe with modern haft (Museum of London)

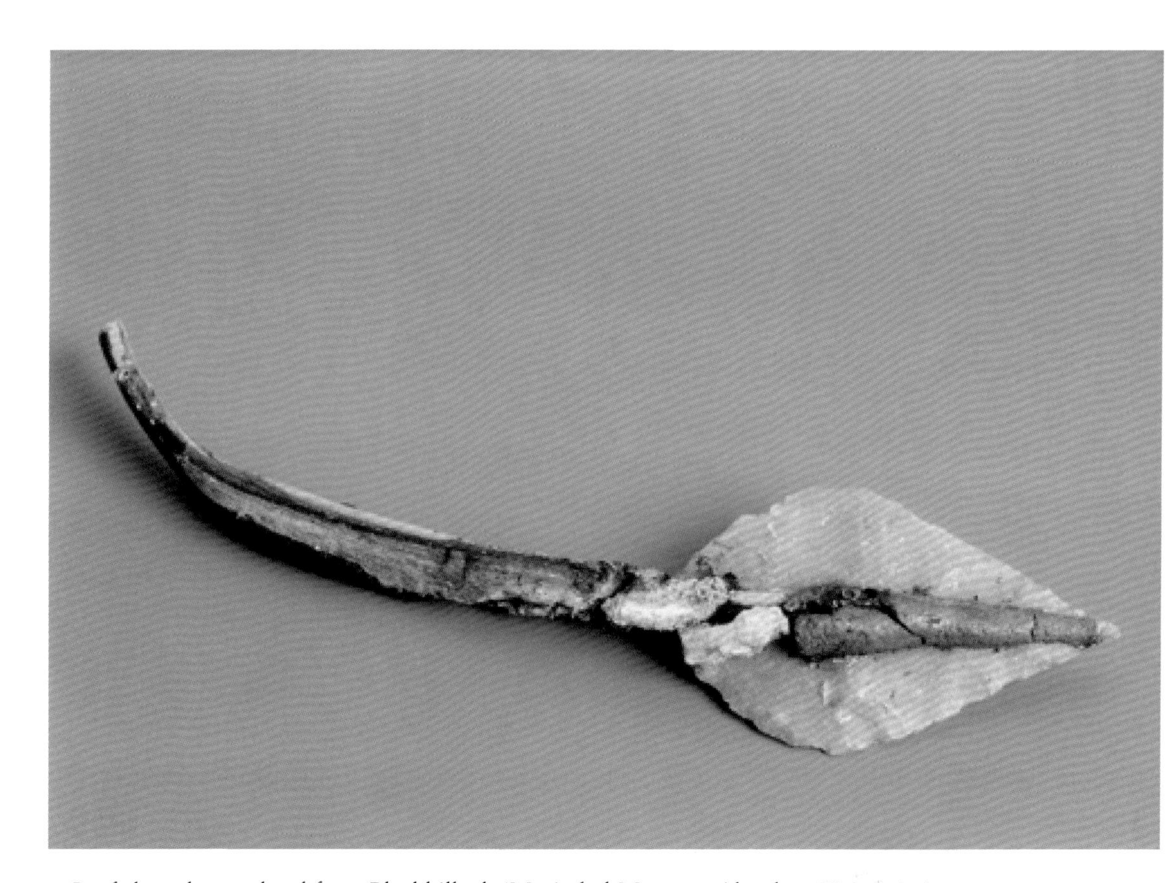

4 Leaf-shaped arrowhead from Blackhillock (Marischal Museum, Aberdeen University)

5 Crickley Hill (Mike Garner)

6 Leaf-shaped arrowheads from Carn Brea (Royal Cornwall Museum)

7 Reconstruction of Whitehorse Stone longhouse (Oxford Archaeology)

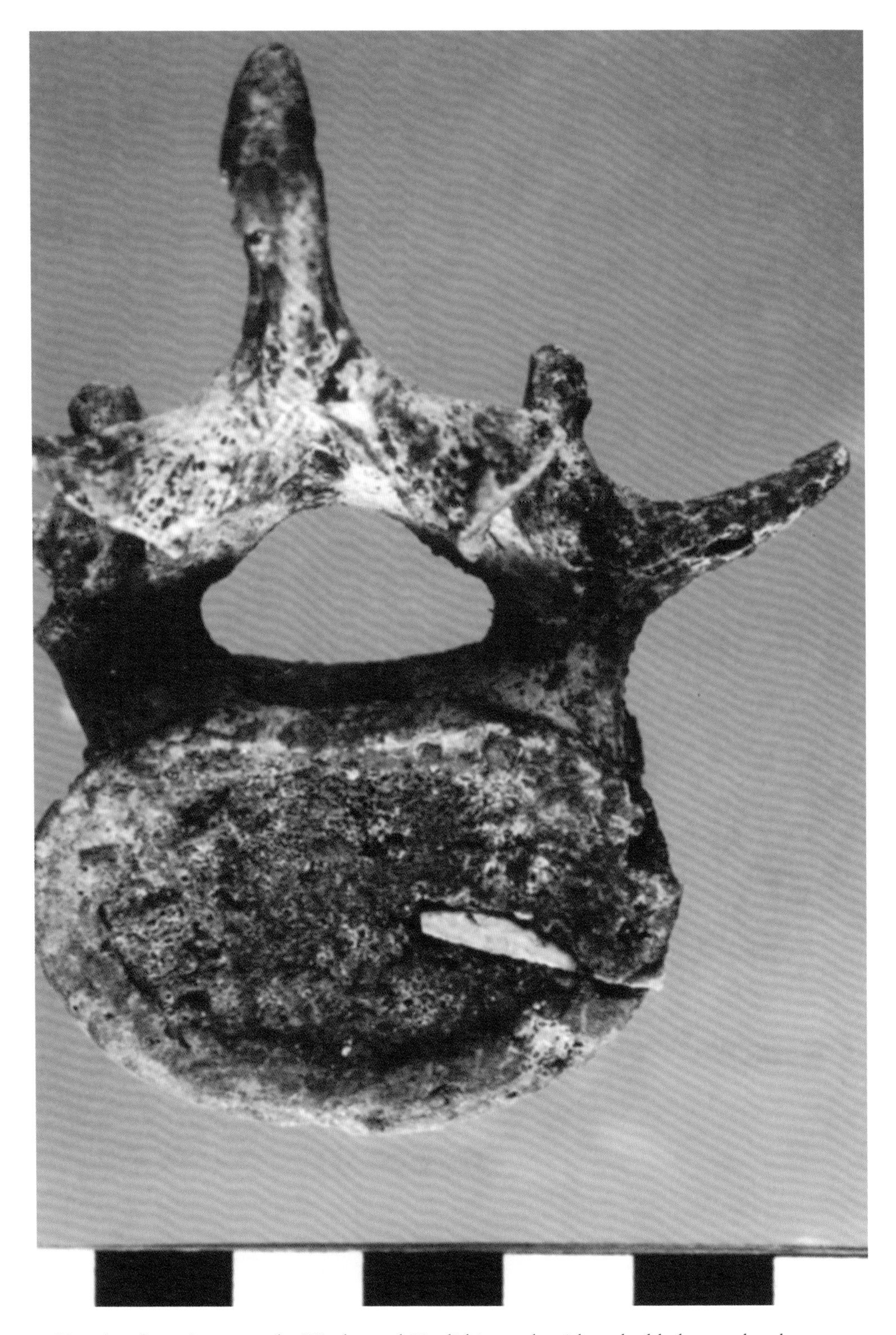

8 Vertebra from Ascott-under-Wychwood Neolithic tomb with embedded arrowhead (Don Benson)

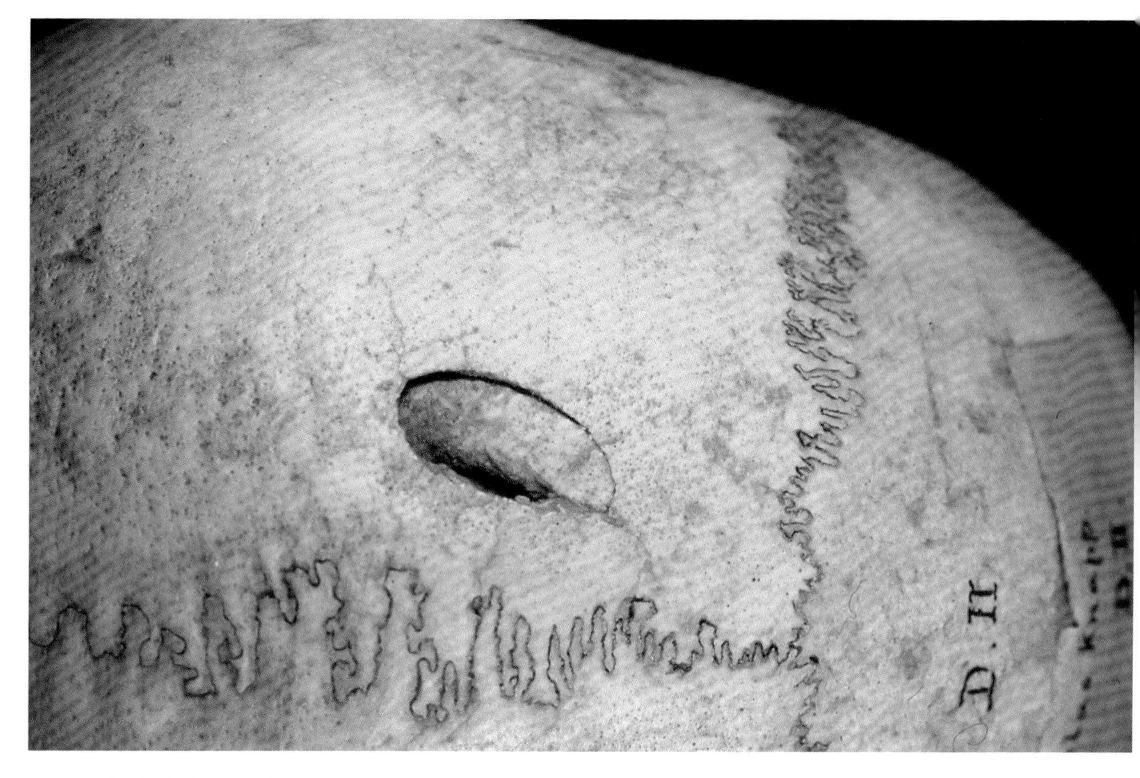

9 Female skull from Belas Knap Neolithic tomb with unhealed depressed oval fracture (Duckworth Collection, Cambridge/Rick Schulting)

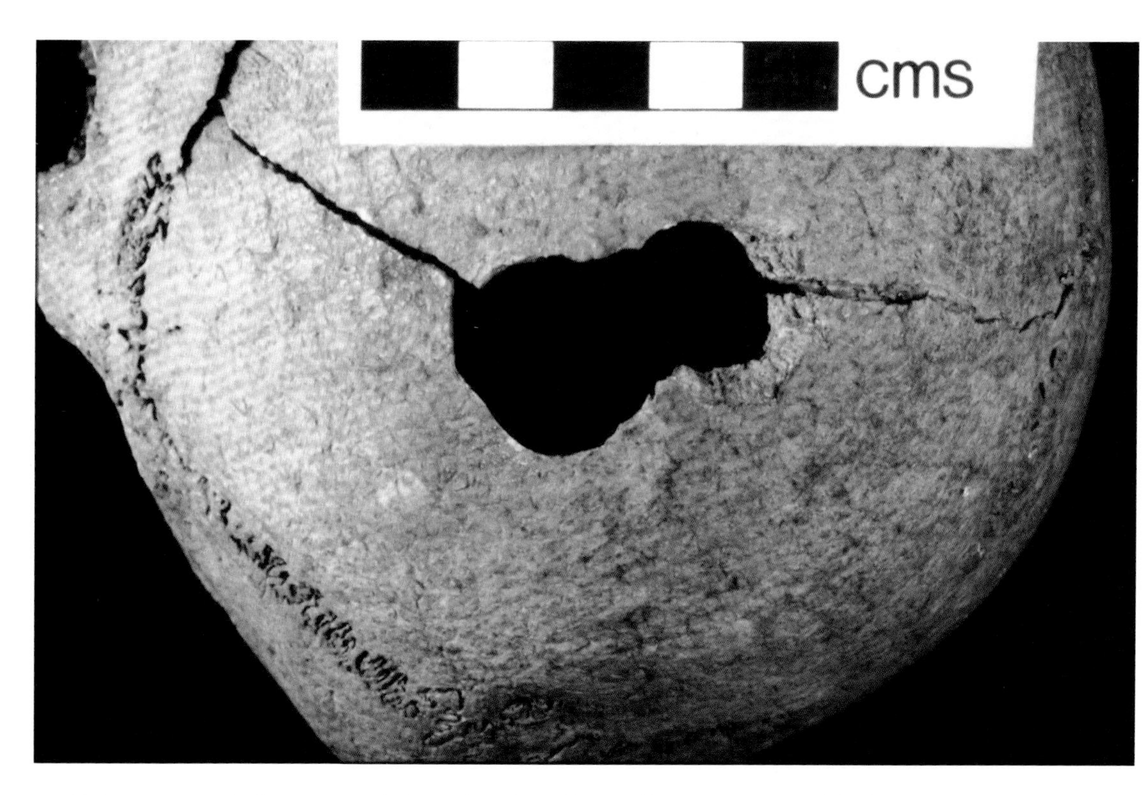

10 Unhealed keyhole fracture on skull (male?) from Dorsetshire Long Barrow (Duckworth Collection, Cambridge/Rick Schulting)

11 Female skull from Dinnington Neolithic tomb with unhealed depressed fracture (Natural History Museum/Rick Schulting)

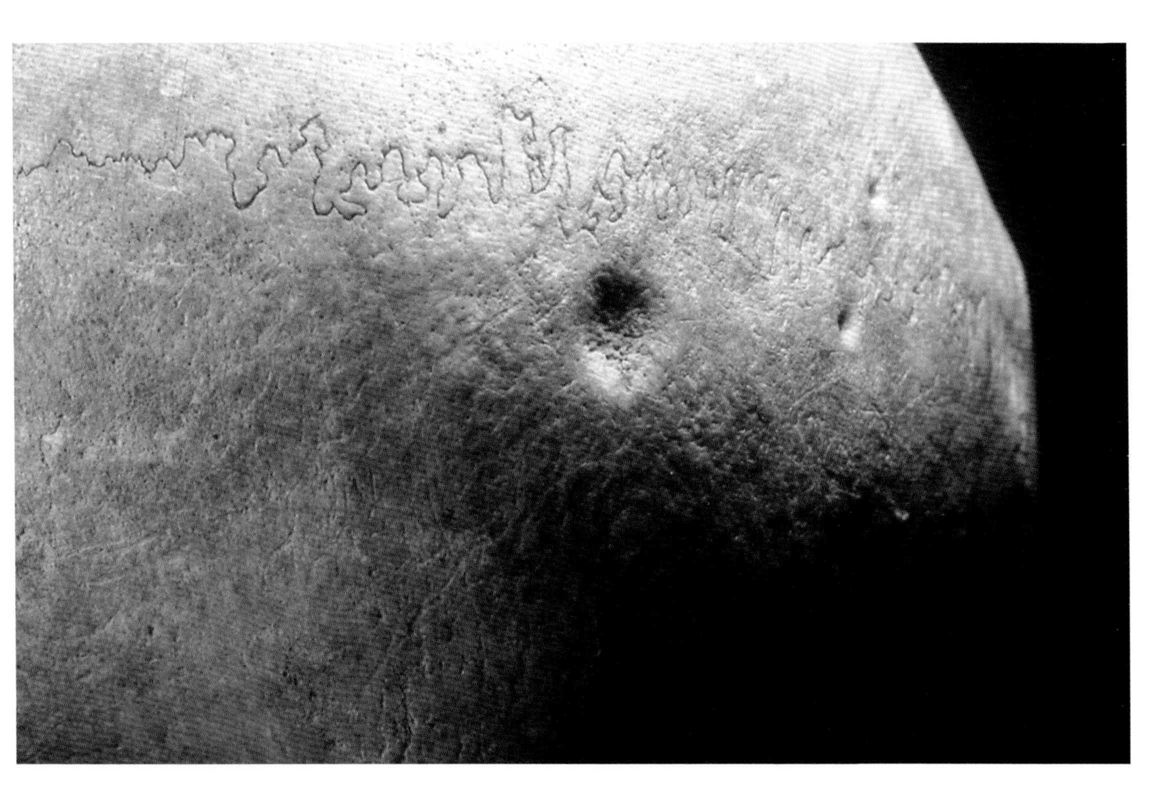

12 Female skull from Dinnington Neolithic tomb with healed depressed fracture (Natural History Museum/Rick Schulting)

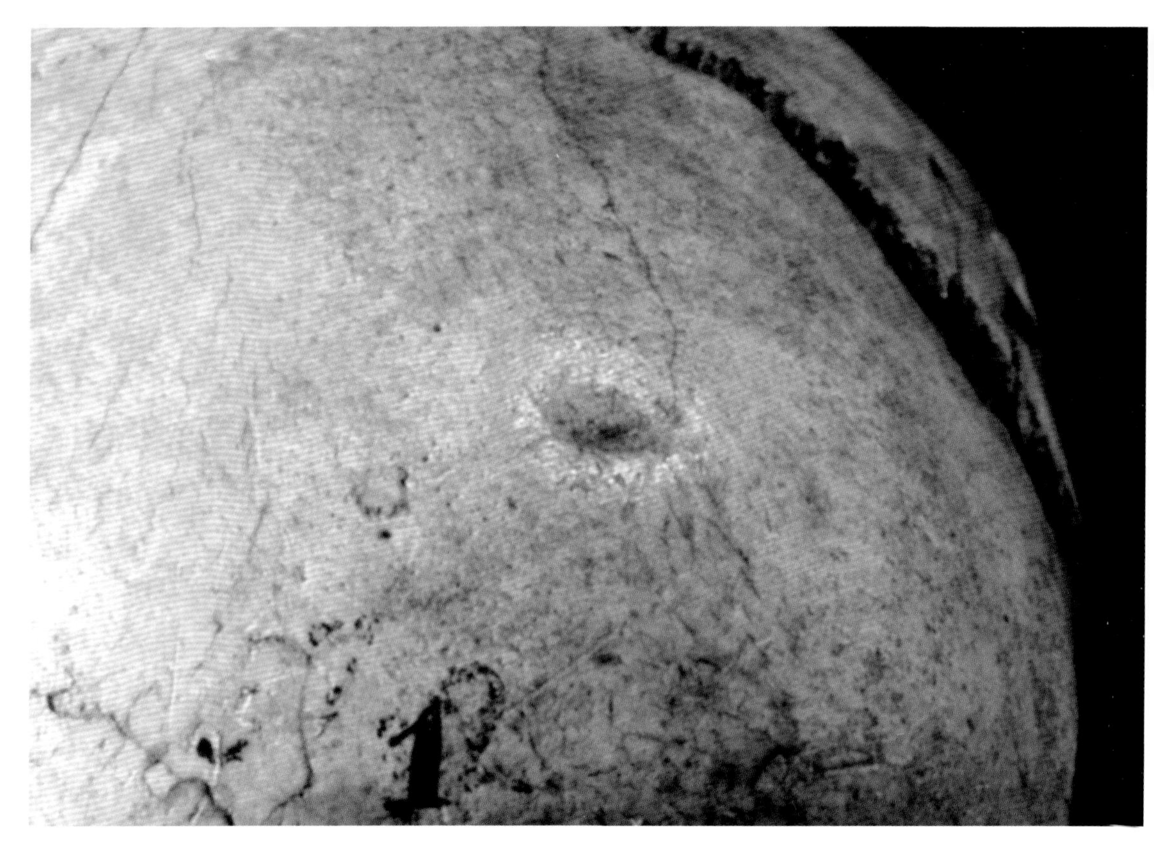

13 Male skull from Norton Bavant Neolithic tomb with healed depressed fracture (Duckworth Collection, Cambridge/Rick Schulting)

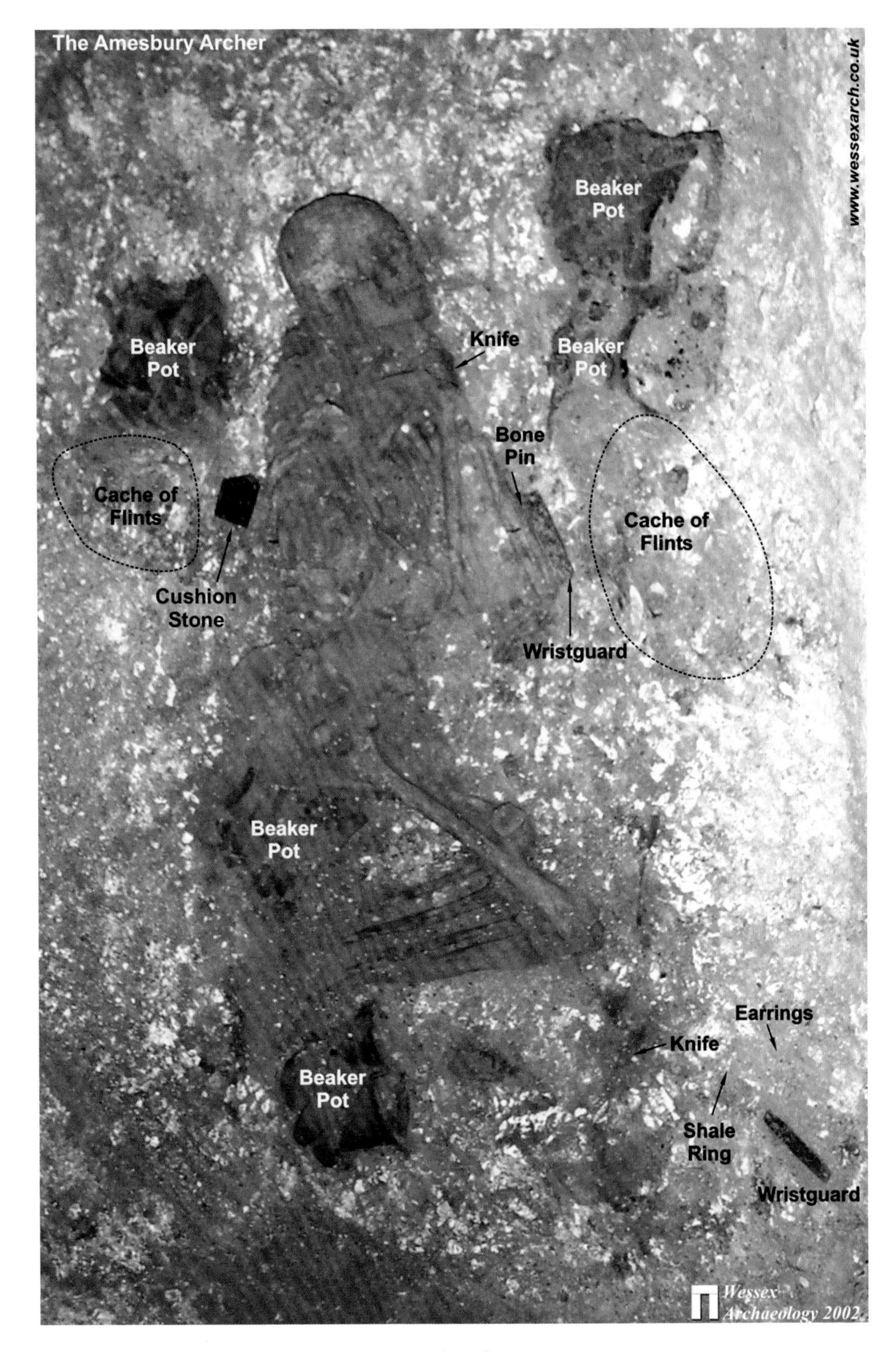

14 The Amesbury Archer's grave (Wessex Archaeology)

15 Amesbury Archer's bracers (Wessex Archaeology)

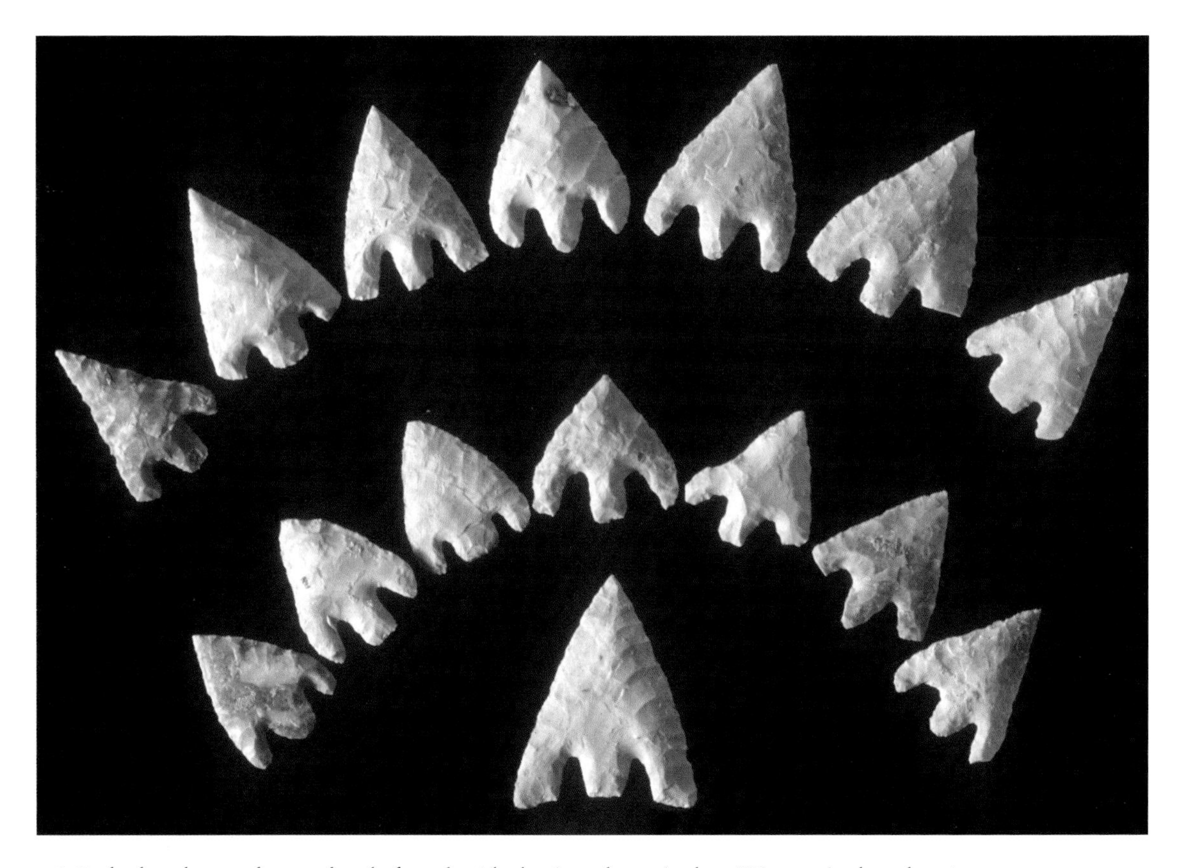

16 Barbed and tanged arrowheads found with the Amesbury Archer (Wessex Archaeology)

17 Reconstruction of early Beaker found with the Amesbury Archer (Wessex Archaeology)

18 Copper daggers found with the Amesbury Archer (Wessex Archaeology)

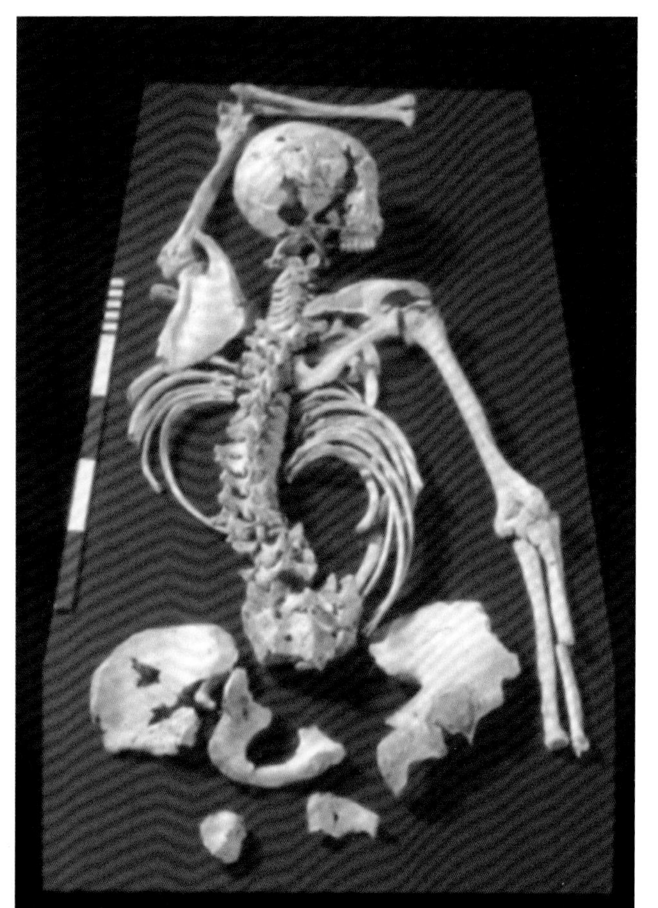

Above: *19* Arrowhead in the spine of a male Beaker burial from Barrow Hills (Oxford Archaeology)

Left: *20* Skeleton of younger male from Tormarton – a probable spear-hole is just visible on side of skull (Richard Osgood)

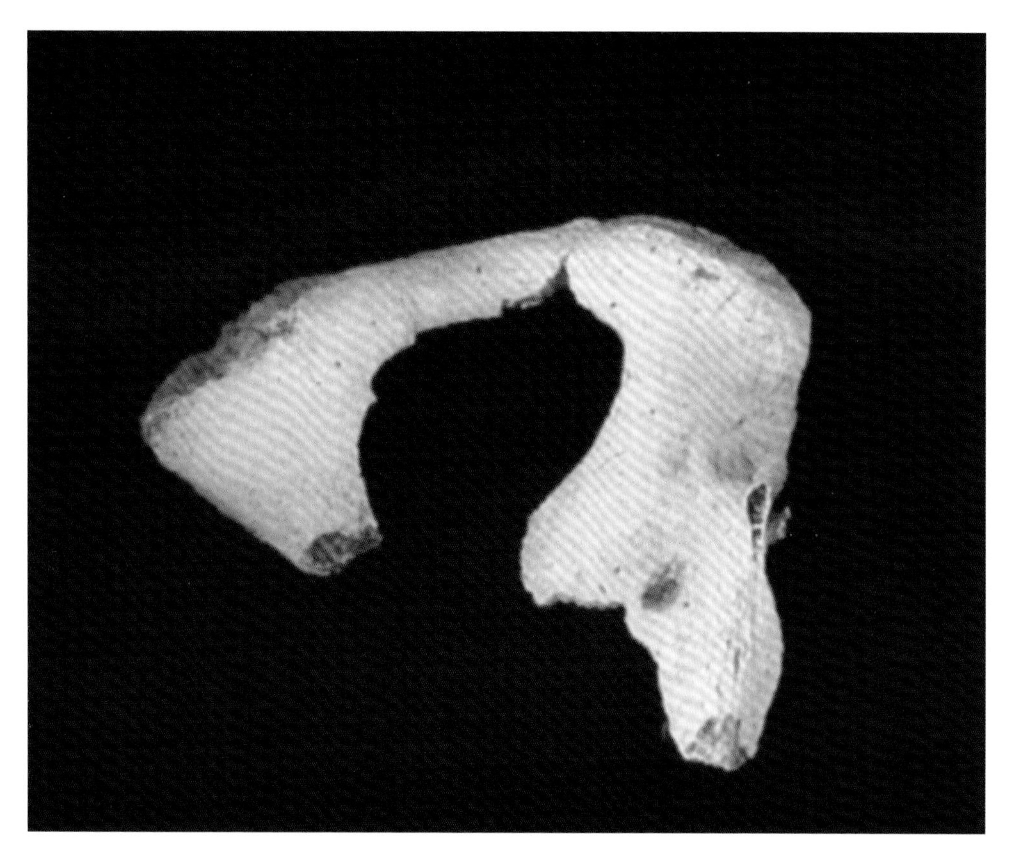

21 Pelvis of younger male from Tormarton with embedded spearhead (Richard Osgood)

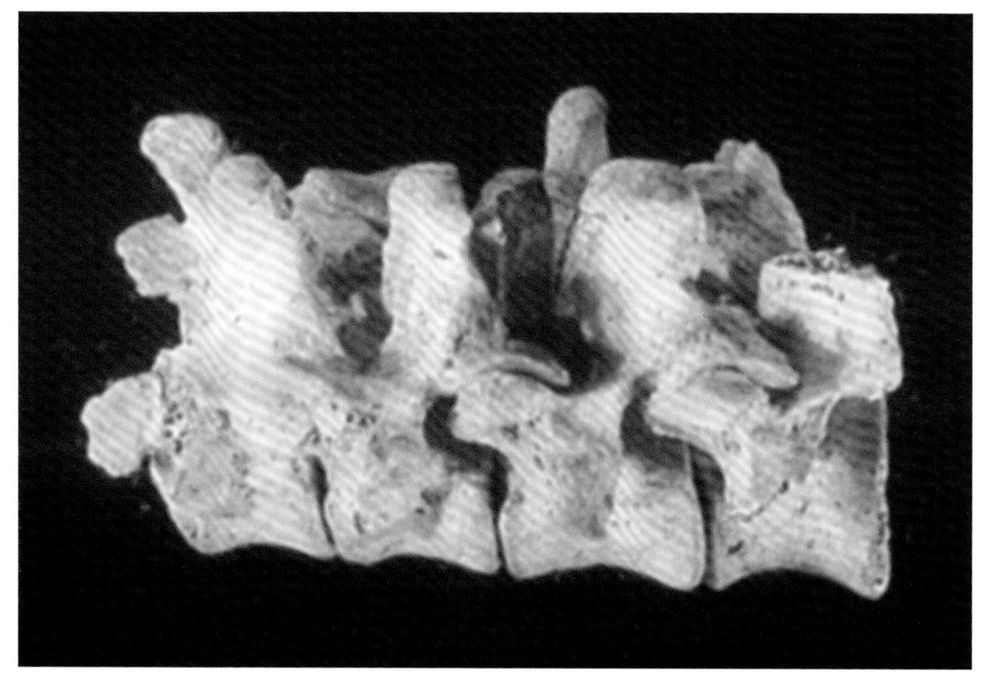

22 Spine of younger male with embedded spearhead from Tormarton (Richard Osgood)

23 Spearhead from spine of younger male at Tormarton (Richard Osgood)

24 Leaf-shaped swords of the Late Bronze Age (Cheltenham Art Gallery & Museum)

Top: 25 Ramparts of Iron Age hillfort at Hambledon Hill (Marilyn Peddle)

Above: 26 The Kirkburn Sword

Right: 27 The Battersea Shield (British Museum)

28 Lindow Man (British Museum)

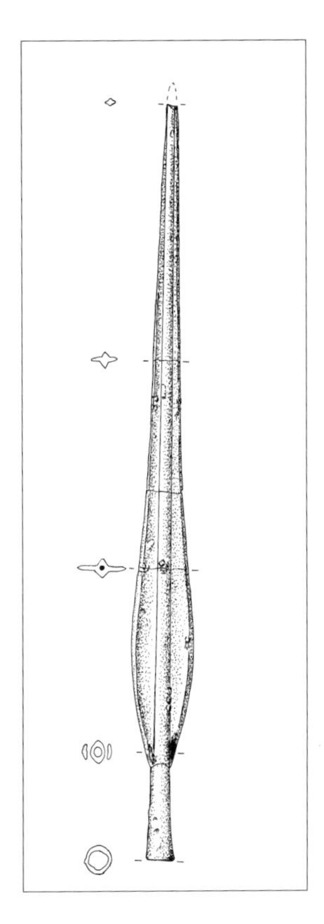

63 Ceremonial spearhead from Croydon (redrawn after Needham)

were a deliberate tactical innovation of Later Bronze Age warfare, Bartlett and Hawkes' theory remains an interesting one nonetheless.

It should be borne in mind that whilst many spearheads used in the Later Bronze Age in Britain probably saw action in combat, the size and general unwieldiness of some examples indicates that they were made purely as ceremonial or parade items. The spearhead (over 80cm long) found at Wandle Park in Croydon probably provides us with one such example (*63*) (Needham 1990) and examples of ceremonial spears can be seen in the ethnographic record. For example, the Babur people of northern Nigeria had five special spears which were used only in rituals or ceremonies and as battle standards that were carried at the forefront of their army (Spring in Osgood 1998: 117)

Another deadly innovation of the Later Bronze Age was the development of the sword and as history clearly shows, this weapon went on to have a long and bloody career in warfare. Although it is not unlikely that some swords were used for 'ceremonial fights or display' (Harding 1999: 167) in the Later Bronze Age, it is inconceivable that they were not used in warfare and swords were probably responsible for many deaths. The swords of the Late Bronze Age were

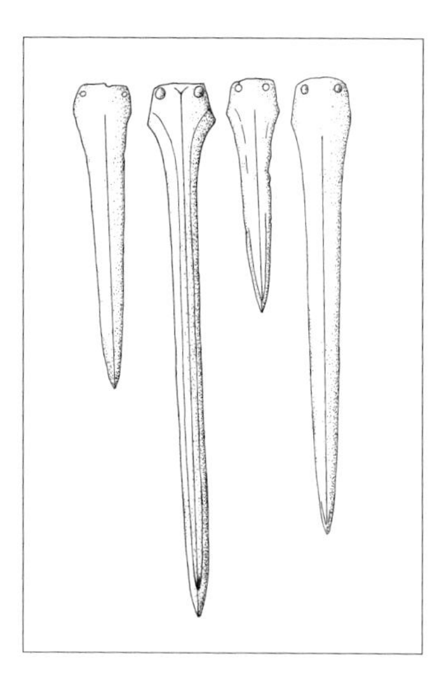

64 Dirks and rapiers of the Middle Bronze Age (redrawn after Trump)

preceded by the dirks and rapiers of the Middle Bronze Age, *c.*1400-1100 BC (64). Whilst these weapons are not generally classed as 'true' swords, because of their smaller size and narrower blades, they would have possessed a much more lethal potential than the Early Bronze Age daggers from which they developed (Thorpe 2006: 155).

Although dirks and rapiers seem to have been primarily designed for thrusting or stabbing, blade-on-blade damage and torn rivet-holes seen on examples examined as part of Jill York's analysis indicate that like later swords they were also used as slashing weapons. That rapiers were used in this manner is perhaps hardly surprising, as it has been plausibly argued (Clements in Molloy 2008: 124) that it very rare to actually find a sword from any period that was made purely as either a cutting or thrusting weapon. A similar logic can be applied to the dirks, as they could certainly have caused vicious injuries if used as a slashing weapon.

Following Rowlands, Roger Mercer (2006: 142) has suggested that as with the rapiers used by the honour-bound societies of sixteenth-century medieval Europe, those of the Bronze Age may have been weapons that had more to do with defending oneself from personal assault or assassination, rather than weapons used in combat by larger groups of fighting men. It is an interesting idea that is certainly worth bearing in mind, although as Mercer (*ibid.*) himself admits, there may have been occasions when 'rapier-girt' leaders led their men in acts of brutality such as that witnessed at Tormarton, although he stops short of calling such acts actual 'warfare'. Richard Osgood (2000: 32) has followed a similar train of thought to Mercer and has suggested that rapiers might have been used in encounters similar to the one between Paris and Menelaus famously

65 Duelling individuals depicted on Mycenaen gold seal-ring

depicted in Homer's *Iliad*, with 'champions' or 'heroes' from opposing warrior elites fighting to the death with their weapons. Similar combat can be seen in the Irish Early Medieval sagas, which contain echoes of the reality of Iron Age (and probably Bronze Age) life: 'when the contending parties were ready for battle, the most prominent warriors from one of the parties stepped forward and challenged to single combat the most distinguished warriors from the other party' (Petterson in Kristiansen 2002: 329). In fact, we do have pictorial hints which indicate that in the Aegean at least, such 'duels' may indeed have been a feature of Later Bronze Age life in Europe, as depictions of combatants duelling with rapiers and shields can be seen on rings, gems and seals made by the fascinating Minoan and Mycenaean societies of the Later Bronze Age (*65*).

It is also apparent that as with spears, some swords were ceremonial in nature and were made purely as votive objects to be offered to the shadowy deities worshipped by Britain's Later Bronze Age communities. Likely examples of such weapons are the superbly made dirk discovered near Oxborough Hall in Norfolk (the blade had been vertically inserted, point downwards into the ground) and the rapier found on a building site in Essex or Kent (Needham 1990). The fact that both blades lacked rivets for the attachment of hilts and had blunt edges lends support to the above idea.

Within a couple of hundred years or so, towards the end of the Middle Bronze Age, dirks and rapiers began to be replaced by 'true' swords. This was probably a natural development borne out of the desire of warriors to have a more versatile and stronger weapon that could be used to cut, parry and thrust from many different angles during close quarter combat (Harding 2000: 277). It is unlikely that such swords were the normal weapons of foot soldiers, or 'infantry'

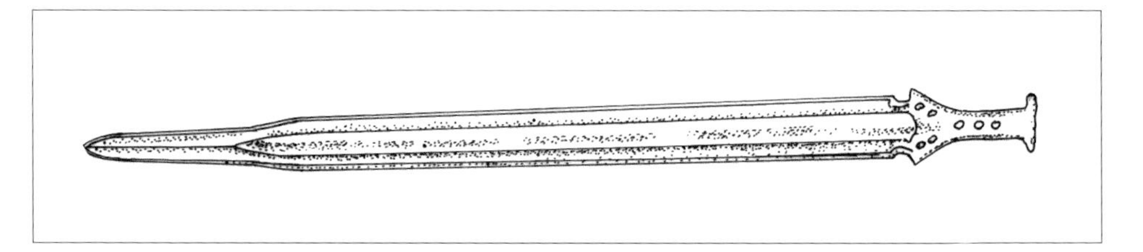

66 Carp's tongue sword (redrawn after Megaw & Simpson)

and it is more probable that they were prestigious weapons wielded by 'chiefly commanders' (Kristiansen 2002: 324). However, the use of swords may not have been restricted to single individuals in Bronze Age war parties, and it is quite possible that several warriors in such groups carried swords into battle.

For the most part, swords of the Late Bronze Age had leaf-shaped blades (*colour plate* 24), although there was considerable variation to be seen in their hilts, with some made from bronze and others clad in bone or wood plates (Osgood 2000: 24). At the end of the Late Bronze Age, the so-called 'Carp's Tongue' sword (*66*) became dominant in Britain and the Atlantic zone of Europe in general, and these weapons appear to be a deliberate amalgamation of rapiers and leaf-shaped swords in order to produce an even more deadly close-combat weapon (*ibid.*). As with all Later Bronze Age weaponry in Britain, there can be little doubt that some of the swords from this time had more to do with social display rather than actual warfare and some could have been made purely as offerings to the gods. However, a revealing study of Late Bronze Age swords from Ireland showed that over 90 per cent of the swords displayed edge damage which is likely to indicate their use as weapons (Bridgford in Osgood 1998: 13). Likewise, Kristian Kristiansen (2002: 323) has also commented on the fact that on many Bronze Age Swords, 'the blade below the hilt is the area of defence and here one often finds severe damage and resharpening'. Whilst it is unlikely that all of this damage occurred in warfare, it is probable that a large amount of it did occur through the use of the swords in combat.

An intriguing discovery in respect of the use of swords in Later Bronze Age warfare is the wooden sword (*c.*30cm in length) made from yew that was recovered from a peat bog at Lang Hill in Scotland (Stevenson 1960). In addition to a broken tip, the hilt of the sword had been polished by repeated handling, which indicates that it had been used for some time before it was probably ritually deposited in the bog. Although the sword may actually have been used as weapon (the use of wooden swords in combat is attested in the Irish sagas), it is perhaps more probable that it was used in the training of apprentice swordsmen (Kristiansen 2002: 325). It is unlikely that warriors equipped with swords would simply go into deadly combat without some form of schooling in the defensive and killing capabilities of their weapons.

DEFENSIVE EQUIPMENT

As we have seen, shields first appeared in Britain (or at least Ireland) during the Early Bronze Age. Whatever the true date of the first British shields, they were unsurprisingly more common in the Later Bronze Age and were clearly a necessary accoutrement for those involved in warfare.

The superbly made bronze shields produced during the Later Bronze Age represent some of the finest artefacts to have come from prehistoric Britain, but were they actually used in warfare? John Coles' early and famous example of experimental archaeology, in which he found that a replica leather shield stood up much better to sword blows than a replica 'bronze' one (the shield was actually made of copper), suggested they were not (Coles 1962). Subsequently, the dominant view has been that bronze shields were too thin to have offered any real protection from swords and spears and that they were made as ritual and ceremonial items. There can be no denying that Coles' experiment, and the fact that many such shields were deliberately deposited in watery places, reveals the true nature of many of the Later Bronze Age shields in Britain. However, as Osgood (1998: 9) has stated, the shields of the *Nipperweise* class were made from cast bronze and were much thicker than other types. Therefore, 'To dismiss the use of bronze shields in battle' (*ibid.*) would indeed be wrong.

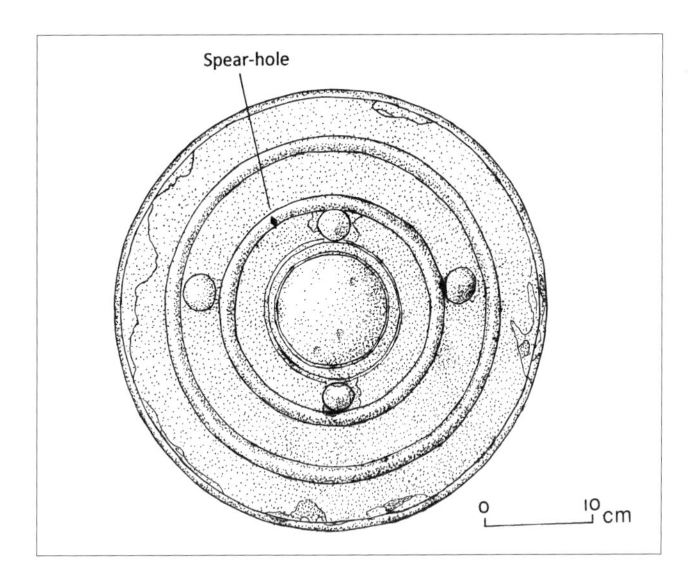

67 The Long Wittenham shield (redrawn after Needham)

In fact, a likely example of a *Nipperweise* shield that was used in actual combat can be seen with the well-known and intriguing example found at Long Wittenham in Oxfordshire (Needham 1979). That this shield had been used in warfare was implied by the fact that it displays four perforations, the largest of which 'has a concave lozenge shape reminiscent of a midrib spearhead section' (67) (*ibid.* 113). In addition to this telling damage, two of the other holes were lozenge shaped and the fact that they were later mended by hammering flat lends considerable weight to the idea that the shield had actually seen combat, at least on a couple of occasions (Osgood 1998: 9).

However, while the Long Wittenham shield was probably damaged in battle, it is far more likely that leather and wooden shields were more commonly used by warriors as they would have offered greater protection. Although no such shields have been found in Britain, we do have rare Irish examples; a leather shield was found at Clonbrin, County Longford and two wooden shields have been recovered from bogs at Annandale, County Leitrim and Cloonlara, County Mayo (Coles 1962: 160).

Considering the abundance of new weapon types in the Later Bronze Age, it would be somewhat strange that armour of some form or other was not sometimes used in conjunction with shields during this time. On the continent, superb examples of decorated sheet bronze armour have been found, such as the famous decorated cuirass from Marmesses in France. However, these examples like others were more than likely used for display rather than as a defence against weaponry, and hardened leather or some other organic material would have offered much greater protection against weapon blows. Nonetheless, it is possible that sheet Bronze armour might in some cases have been worn over heavy padding and appeared on the battlefield because of the psychological effect that it could have on the enemy. As Anthony Harding (2000: 288) has pointed out, 'The mere sight of Achilles' armour, as worn by Patroclus in Book XVI of the Iliad, was enough to strike terror into Trojan hearts'. Whilst such ancient literary evidence cannot of course be wholly relied upon, it provides us with hints as to how metal armour may have been used and perceived in the Bronze Age world. Staying in the world of the Bronze Age Mediterranean, the remarkable and sophisticated suit of bronze armour found in a Mycenaean tomb at Dendra may have been made for use in combat, although its heavy weight would effectively stop its wearer from getting up again if they fell, or were knocked down (Dickinson 1994: 205).

Unfortunately, it appears from the archaeological record, that for some reason, bronze armour was not used in Later Bronze Age Britain, as it was in many parts of continental Europe. However, as the familiar archaeological saying has it, 'absence of evidence is not evidence of absence' and maybe one day, a remarkable find of such armour will come to light.

HILLFORTS AND OTHER FORTIFIED SITES OF THE LATER BRONZE AGE

Although hillforts are often seen as a defining feature of the Iron Age, it is clear that they were also built in the Later Bronze Age and it is likely that many more lie hidden beneath the ramparts of the Iron Age forts that succeeded them. Although there is not space here for a detailed discussion of the many examples that can be found in Britain some of the more notable examples will be briefly examined.

One of the best known hillforts of the Later Bronze Age is that at Rams Hill on the Berkshire Downs, which was investigated by Richard Bradley and Anne Ellison (1975). During their excavations they discovered an enclosure ditch that was filled with a mass of chalk rubble and which was backed by a double line of postholes. In regard to this evidence Ellison and Bradley (*ibid.* 35) plausibly suggest 'that the rubble in the ditch represents the facing of an obsolete defence which was cut back ... to allow a massive and close set palisade to be bedded deeply in the chalk. This might then have acted as a new front revetment to a more impressive fortification'. Stuart Needham and Janet Ambers (1994: 227) have proposed a slightly different sequence for the different phases of the defences at Rams Hill, with the final phase seeing 'the placing of a double palisade

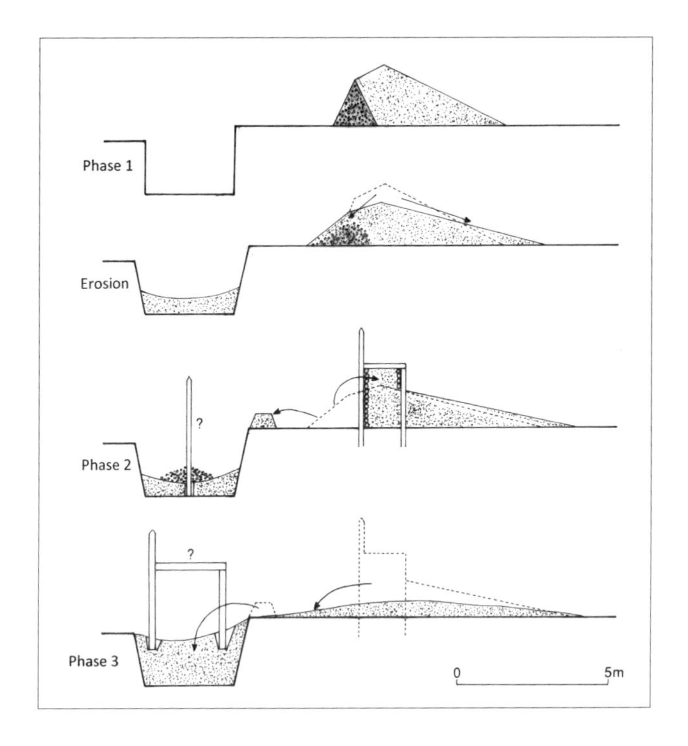

68 Proposed rampart sequence at Ram's Hill (redrawn after Needham & Ambers)

in slots cut into the top of the ditch, now almost fully silted up; the palisades were presumably either tied together with cross-beams or braced independently in some other fashion' (*68*). Radiocarbon dates obtained from various contexts within the defences suggest that they were built in the latter centuries of the second millennium BC (Needham & Ambers 1994: 235; Osgood 1998: 57).

Equally as well known as Rams Hill, is the possible Bronze Age hillfort stunningly situated on the top of Mam Tor (which rises to around 510m) in Derbyshire. Today the site consists of a single dump rampart from the Iron Age, which encloses an area of some 16 acres and a single posthole found underneath the rampart may indicate the existence of an earlier palisade enclosure (Osgood1998: 60). Radiocarbon dates obtained from some of the dwellings within the enclosure spanned the period 1679-996 BC and the excavators of the site concluded that initially, this *Höhenseidlung* (elevated settlement) had been defended by a simple palisade which was later replaced with ditches and banks (*ibid.*).

Another possible defended site of the Later Bronze Age is that found at Hog Cliff Hill in Dorset (Ellison & Rahtz 1987). This site consisted of a large oval ditch enclosing an area some 200m wide and 280m in length. However, although it bears some similarities to Rams Hill with its oval enclosure and hilltop location, the fact that its ditch was shallow (an average 0.7m deep) and that it was inside an insubstantial bank argues against a defensive function.

More suggestive evidence in regard to Later Bronze Age fortification of a hilltop site comes from Norton Fitzwarren in Somerset. This site was first identified as a possible Later Bronze Age hillfort by St George Gray who found Middle Bronze Age pottery and a hoard of metalwork associated with a ditch and counterscarp bank (Osgood 1998: 57). Later excavations at Norton Fitzwarren (Ellis 1989) revealed that an outer bank and inner ditch had enclosed the site in the Middle Bronze Age, but that this may have been replaced by a wooden palisade at a later date. Thus it is quite possible that a ceremonial enclosure was later transformed into a defended one and the production of swords that is known to have taken place at Norton Fitzwarren (as evidenced by the clay sword mould fragments found here) could well be linked to this transformation (*ibid.*: 66).

Turning from English examples of Later Bronze Age hillforts to Welsh ones, there are the notable sites of Dinorben and the Breidden. The former lies above the Vale of Clwyd, near the town of Abegerle in North Wales and excavations carried out during the late 1960s revealed the remains of a Late Bronze Age hillfort (dating to around the turn of the first millennium BC) succeeded by an Iron Age example (Savory 1971). The site appears to have been defended by a succession of ramparts, with the Late Bronze Age one consisting of 'layers of rubble or clay staged with rafts and revetted in front with a palisade bedded in a trench and at the

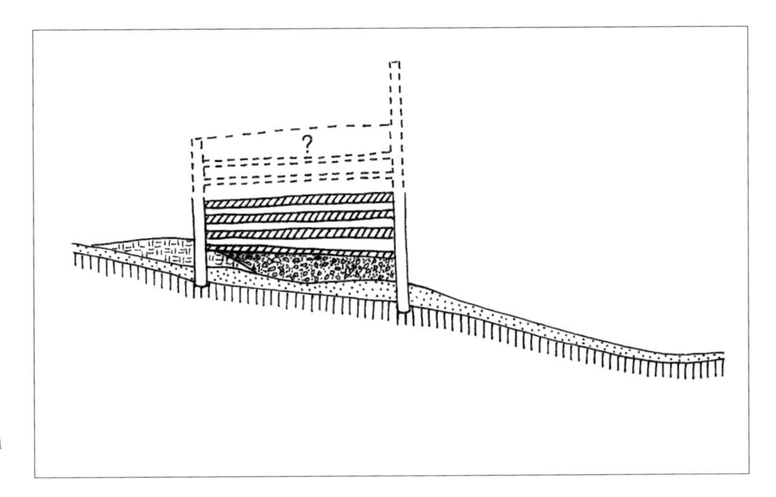

69 Rampart at Dinorben (redrawn after Savory)

back, probably, with horizontal planks fastened to widely-spaced uprights' (Savory in Osgood 1998: 64). In other words, the defences at Dinorben during the Late Bronze Age probably consisted of a box rampart (*69*). The building of box type ramparts may reveal the spread of ideas (or perhaps even people in some cases) from the Late Bronze Age Urnfield Culture of central Europe (Osgood 1998: 64).

At the Breiddin (Powys) a substantial rampart was built around a Late Bronze Age settlement and associated radiocarbon dates ranged from around 1050-800 BC (Harding 2000: 298). The rampart was again of the timber-framed box type and it had a rubble filled core that contained numerous stones of varying size (Osgood 2000: 13). Evidence found elsewhere in this area of the Welsh Marches suggest that like the Breidden, a number of Iron Age hillforts (e.g. Moel y Gaer) can trace their beginnings back to the Late Bronze Age (Harding 2008: 298).

Interestingly, both Dinorben and Breiddin have provided further evidence which indicates that they were attacked during the period of their Late Bronze Age occupancies. At Dinorben, the remains of three males were discovered in the ditch bottom and the skull of one of these seems to have been cut in two, whilst at the Breidden, the box-rampart may have been destroyed by fire in an attack on the site (Gardner & Savory; Musson: in Osgood 2006: 158).

Another intriguing Welsh example of a likely Late Bronze Age fortified settlement is Castell Odo (Mynydd Ystum), which is situated near the tip of the Llŷn Peninsula in north-west Wales. This multi phase site was first excavated by C.E. Breeze (1932) and ceramics found during his work led to the long-held belief that Castell Odo was a classic Dark Age fortified site of the post-Roman period. However it is now recognised that this pottery is actually Late Bronze Age in date (Lynch 2000: 151) and that the first phase of the site dates to this time. Further excavations were conducted by Leslie Alcock (1960) in 1958-59 and he found evidence that the Late Bronze Age settlement had probably been attacked, with the remains of burnt wooden buildings found as well as the probable remnants of an unfinished wooden palisade that had also been destroyed by fire. Although this evidence could of course represent an unfortunate accident, it is hard to disagree with Alcock's reading of it; 'In a moment of external danger it had been decided to fortify Mynydd Ystum with a timber stockade ... Its defences unfinished, the settlement was readily stormed. Its houses were burnt down and the palisade ... was destroyed' (*ibid.*: 89-90).

In Wessex there are the *ridge-end hillforts*, which appear to represent defended sites, as their characteristic features are 'strong fortifications and a dominant position, usually on a spur, visible for miles around' (Cunliffe 2004a: 72). Examples of these forts include Giant's Grave, Lidbury Camp, Oliver's Camp and Winklebury (*ibid.*)

In addition to the hillforts of the Later Bronze Age, there are the sites known as 'ringworks'. Although they are mainly confined to the Thames Valley and south-east England, they are found in other areas of Britain and stone versions were probably also constructed in Ireland (Thorpe 2006: 157). It is probable that some sites had a non-military function and were used as religious centres, enclosures to keep animals in or out, or simply as places that reflected the high status of their builders (Osgood 2000: 14-15). However, with others, the term 'ringfort' seems more appropriate as evidence found at some examples suggests that they were designed with defence in mind.

A probable candidate in this respect is the site of Thwing in the eastern part of the Yorkshire Wolds (Manby 2007). Here, the evidence uncovered indicated that a very large timber building (perhaps, but not definitely a shrine) had been surrounded by an enclosure (*c.*130m in diameter) which comprised of a double post setting which supported a chalk rampart that was revetted with timber on its front. Two opposed timber gateways gave access into the interior of the 'fort' and the huge main posts of these may have carried towers with walk-ways over the roadways. In front of the rampart there was a substantial steep-sided enclosure ditch measuring some 4m wide and 2.5m deep.

Another likely ringfort was located at what is now the site of Queen Mary's Hospital at Carlshalton in Surrey (Adkins & Needham 1985). At this site, numerous Late Bronze Age artefacts were found in the substantial V-shaped enclosure ditch that measured *c.*3.5m wide, 2m deep and enclosed an area some 150m in diameter. Although no traces of a rampart were found, chalk blocks found in the ditch may represent the facing of one that had originally stood inside the ditch.

It is apparent then, that in the Later Bronze Age it became increasingly common for people to build fortified sites in naturally defensible positions in the landscape. It seems likely that the protection of vulnerable farmland (which was probably growing scarcer because of population growth and an increasingly poor climate) and a build up of 'wealth' in the form of agricultural surplus and trade goods of both an everyday and prestigious nature, had a large part to play in their appearance (Osgood 2000: 34). Such wealth is likely to have provided the impetus for an increase in raiding and aggressive competition between communities (*ibid.*). It has also been suggested (Harding in Thorpe 2006: 158) that the theft of cattle for prestige could have been one of the motivating factors behind these raids. The fact that many fortified sites of the Later Bronze Age seem to be associated with field systems and are located in positions which dominate natural trade routes such as rivers and passes through higher ground, lend some credibility to the above ideas.

HORSE-BORNE AND RIVERINE RAIDERS?

The fortified sites of the Later Bronze Age surely reveal that during this period there was a very real need for defence and that during this time larger scale raids were becoming increasingly common. Whilst there can be little doubt that some raids were undertaken by war bands travelling under their own steam on foot, there are clues found in the archaeological record which perhaps indicate that other means of transport were used by raiding parties.

One possibility is that warriors used boats to carry them close to enemy sites they intended to raid, or to suitable ambush points on trade routes. Rare but remarkable examples of planked boats and logboats/dugout canoes dating to the Late Bronze Age have been found in various parts of Britain (McGrail

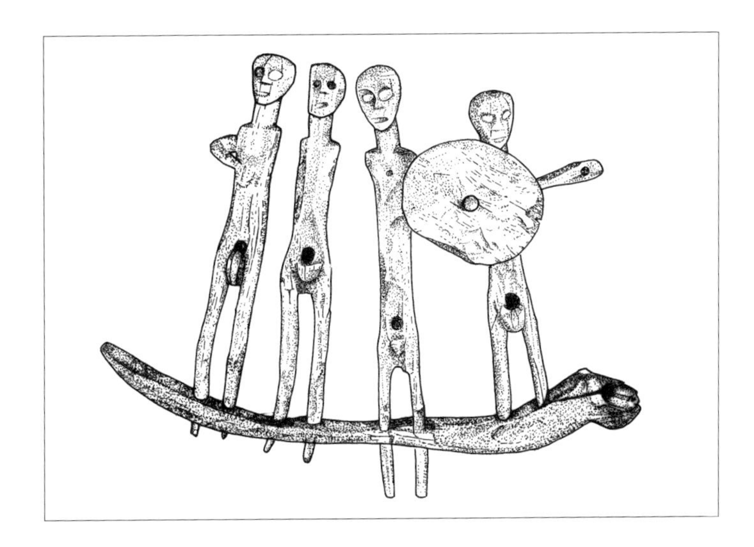

70 Roos Carr figures

1979). A remarkable wooden model found at Roos Carr in Humberside may even provide us with a representation of a Late Bronze Age water-borne raiding party who originally appear to have held round shields and clubs (*70*). Although radiocarbon dates suggest either a Late Bronze Age or Early Iron Age date for the Roos Carr figures, the round shields are more suggestive of the Late Bronze Age rather than the Iron Age when rectangular shields were the norm (Osgood 1998: 40). It is interesting to note that in southeast Norway and Bohüslan many Middle Bronze Age rock art scenes show warriors in boats who appear to be fighting with swords and axes (Fyllingen 2006: 321-322). Such depictions provide us with strong hints that boats were an important means of transportation for those engaged in warfare.

In addition to boats, it is possible that horses were used by war bands in Later Bronze Age Britain. Although we have no firm evidence for this practice, horse gear and horse bones found at Late Bronze Age sites reveal that horse power was certainly being harnessed during this time. Sheet bronze discs or 'phalerae' are often found with other items of horse gear and they were probably used to decorate horse straps and harnesses (Osgood 1995: 57). Interestingly, two of the three phalerae that were dredged from the River Avon near Melksham in Wiltshire display perforations caused by weapons; those on phalerae 1 were made by a dirk or rapier (*71*) while phalerae 2 was stabbed with a spear (*ibid.*: 52) The possible significance of this weapons damage will be briefly considered below.

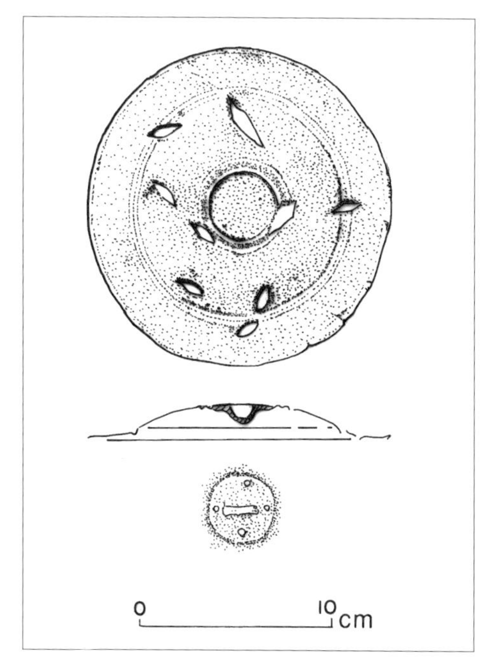

71 Phalerae from Melksham with damage caused by a dirk (redrawn after Osgood)

Although the presence of actual cavalry in the Later Bronze Age (as envisaged by earlier archaeologists) cannot be totally discounted, it is perhaps more likely that if indeed horses were ridden by armed warriors, they were used for mobility rather than as fighting platforms (Osgood 1998: 83). This is suggested by the fact that the horses of this time would have been little bigger than today's Exmoor ponies and that we have no evidence for stirrups, the absence of which would have made fighting on horseback very difficult (*ibid.*). However, mounted warfare would not have been impossible, and of course, stirrups could quite easily have been made from organic materials, as shown by the famous Comanche Indians, who used horsehair bridles, hide saddles and leather stirrups (Burl 1987: 108).

LATER BRONZE AGE HOARDS: AN INDICATION OF WARFARE?

A ubiquitous feature of the Later Bronze Age is the hoards of metalwork that were deposited in rivers. As is the case elsewhere in Europe, Later Bronze Age hoards in Britain vary greatly in size and composition and include tools, weapons, ornaments, ritual objects, personal toilet items and also collections of scrap metalwork consisting of broken objects and casting debris. However, it is clear that in Britain, as in other parts of western Europe, the deposition of weapons took precedence over other artefacts (Bradley 1998: 110) and rivers were the favoured location for their disposal, with the Thames providing us with a notable example of this practice (York 2002). Of course these hoards of weaponry are surrounded by ambiguity and there may have been several reasons why they were consigned to rivers and other watery places (and in several cases, the ground). Nevertheless, there is a strong possibility that in one way or another, they bear witness to warfare. In this respect, Kristian Kristiansen (2002: 329) has noted that the majority of Bronze Age swords deposited in wetland hoards in east central Europe bear signs of combat damage, and he draws comparisons with the Gallic Iron Age practice mentioned by Caesar (*Gallic Wars* Book 6, 17), where the weapons of the defeated were deposited in sacred parts of the landscape to thank the gods for victory. Likewise, Richard Osgood (1998: 14) has asked in regard to the phalerae from Melksham: 'could these be trophies from a skirmish or battle which have been ritually destroyed by weapons, to symbolise the previous owner's defeat in combat, and then deposited in the River Avon to give thanks to the deity responsible for the success of the warrior or tribal grouping?'. A similar tack is taken by York (2002: 90) who has suggested that some of the weapons examined in her study of metalwork from the Thames and its tributaries 'could also represent triumphal display after defeating an enemy'. It has also been suggested (Savory in Osgood 1998:14) that Late Bronze Age hoards found at Cardiff and Llynfawr represent the booty taken in 'cross-border' raids carried out on hillforts in the west Midlands or western England.

Some scholars have proposed (e.g. Needham & Burgess in Osgood 1998: 17) that the apparent combat damage observable on weapons found in rivers may actually represent the actual occurrence of battles or ritual combat at river crossings or fords. River crossings could also have been the scene of the possible ambushes that were considered above, and thus it is conceivable that some of this damage relates to armed conflicts between 'raiders' and 'traders'.

One interesting theory in regard to the large deposition of weaponry in rivers during the Later Bronze Age has been put forward by Richard Bradley (1998: 139). He has suggested that these depositions represent a type of 'surrogate warfare', with competing groups using them in acts of 'military posturing' in areas that lacked fortified sites, or during periods when they were not being built. Therefore, it might be possible that such 'shows of strength' were used to stake a claim to territory and to make a statement about ownership and control (York 2002: 90).

SKELETAL EVIDENCE

We have already seen at Tormarton how spears were used as brutal weapons of war in the Later Bronze Age and archaeological evidence found at Queenford Farm near Dorchester-on-Thames points in a similar direction (Knight *et. al.* 1972). A human skeleton was discovered here in 1901 and it was evident from the snapped bronze spearhead in his/her pelvis that as at Tormarton, the spear had been thrust into the body with great force and had broken when being withdrawn. The skeleton was dated to *c.*1260-990 BC, which revealed that this individual's life was cut violently short in the Later Bronze Age.

A discovery from Ireland can probably be seen in the same grim light as that found at Tormarton and Dorchester-on-Thames. This consists of a socketed dagger of the Late Bronze Age, which was embedded in the skull of a skeleton discovered by workmen in the 1830s at Drumman More Lake in County Armagh (Waddell 1984: 71).

At the fascinating and long-lived site of the 'Sculptor's Cave' on the south shore of the Moray Firth (Benton 1931), evidence hinting at similar practices to those carried out at such sites as Ofnet Cave has been found. Among the many artefacts and human bones dating from the Late Bronze to the Viking period discovered in the cave, there were nine human cervical vertebrae (some belonging to children) which revealed that some people had been skilfully decapitated Benton 1931: 206 Shepherd 2007: 198). It is possible that skulls were displayed in the entrance to the cave and were suspended from the walls or ceiling (*ibid.*: 199; Shepherd in Brück 1995: 2760). Although there is some uncertainty as to whether all this skeletal evidence actually dates to the Later Bronze Age (Thorpe 2006: 156; Armit 2007: 2), its association with metalwork of this period (Shepherd 2007: 195, 201) suggests that it much of it does. It is likely that such practices relate to some type of 'head cult' at the Sculptor's Cave in the Later Bronze Age and it is not beyond the bounds of possibility that members of this cult used heads that were taken from slain members of an enemy community.

Joanna Brück's (1995) survey of Late Bronze Age human remains in Britain has highlighted a number of examples of skulls, skull fragments and other human bones found in various contexts and it quite possible that that some of this skeletal material is related to warfare. Examples recorded by Brück include part of a skull found at Billingborough (Lincs.) that had been worked into a vessel and the disarticulated and scattered remains of two individuals found under the ramparts of the hillfort at Breedon-on-the-Hill in Leicestershire.

Also of possible interest are the many hundreds of human skulls that were dredged from the Thames during the last century. The majority of the skulls were found in an area that also contained major collections of Bronze Age weaponry and radiocarbon dating carried out on some of these skulls placed them in the Later Bronze Age; intriguingly, many of them belonged to males aged from 25-35 (Osgood 1998: 18; Bradley 1998: 108-109). Is it possible then, that some skulls represent the heads of warriors taken in battle and deposited in the Thames as an offering to the deity or deities associated with the river? Evidence that this practice of skull deposition in the Thames had a long heritage was provided by a Neolithic date obtained from one skull (Bradley & Gordon 1988: 508) and it is quite likely that other undated samples also belong to this period (Knüsel & Carr 1994: 163).

Finally, a male human skull dated to 1040-810 BC was discovered in association with a 'burnt mound' (the true purpose of these intriguing monuments is unclear, but they may have been cooking places or ritual sites), located near the River Soar in Leicestershire (Beamish & Ripper 2000). Along with a decapitated horse skull and the butchered bones of cattle and aurochs, the remains of another male were found. The dated skull had unhealed cut marks on the atlas vertebrae, which may indicate that he had been killed by being beheaded. It is possible that the other individual also met a violent end.

CHAPTER 5

THE CELTIC IRON AGE C.750 BC–AD 43

Although the Greek and Roman writers of the classical world have left us with much fascinating information about the famous Celts of Iron Age Europe, it should be remembered that often they wrote from the standpoint of a 'civilised' world that was justifying the subjugation and exploitation of a 'barbarian' one. Therefore, writings on the 'fierce' and 'savage' Celts that have come from the pens of authors such as Livy, Pausanias and, most famously Strabo – who tells us that 'The whole race is war mad, high-spirited and quick to battle' – should undoubtedly be treated with caution. Nevertheless, there can be little doubt that this biased portrayal of Celtic society also reveals something of the reality of Iron Age life in both continental Europe and Britain. As Barry Cunliffe (2005: 553) says, 'This generalised picture of the Celt is one repeated many times by classical writers and there is no reason to suppose that the British tribes were in any way different: weapons are not infrequently found, massive fortifications abound and bodies with battle wounds are by no means uncommon'.

IRON AGE HILLFORTS AND THEIR DEFENCES

The hillforts of Iron Age Britain undoubtedly provide us with one of the most distinctive and often awe-inspiring reminders of the prehistoric past and many hundreds of examples still survive throughout Britain. There are strong grounds for claiming that the primary role of many of these fascinating and evocative sites was a defensive one.

Ramparts

Approaching them from a basic level, hillforts consist of one or more defensive circuits of rampart and ditch, with timber, stone and earth used in their construction. Although thousands of years later many hillfort defences are still hugely impressive, today, they often appear as somewhat uniform grass-grown ditches and banks. In recent years however, archaeologists have gone some way to unravelling the history of these fascinating sites. As Barry Cunliffe (*ibid.*: 364) has noted, excavation and a large series of radiocarbon dates are beginning to clarify the sequential development of Iron Age hillfort defences throughout Britain.

It is evident that the earliest Iron Age hillforts continued the traditions of their Late Bronze Age counterparts, as excavations at various sites indicate that earth-filled palisades and box ramparts fronted by ditches continued to be built from the eighth to sixth century BC (*ibid.*: 351). This early phase of rampart building is known as the *Ivinghoe Beacon style* after the rubble-filled box rampart identified at a hillfort in Buckinghamshire, which appears to date to the eight and seventh centuries BC (Dyer 1992: 16). These Early Iron Age defences have also been termed 'wall-and-fill' ramparts (Avery 1986).

Following the *Ivinghoe Beacon style* was the *Hollingbury Camp style* rampart, which simply consisted of the addition of sloping wedge-shaped banks to the back of box ramparts. This would not only provide additional strength and rigidity to the rampart, but would also provide easy access to its breastwork (Cunliffe 2005: 351), where sentries could keep watch, or defenders could rain down missiles on enemy forces if under attack.

72 Tre'r Ceiri (author)

Appearing in conjunction with the *Hollingbury style* around the turn of the sixth century BC is the *Moel y Gaer style*, named after the north Welsh site where it was first identified. Here the rampart consisted of a series of box-like rubble filled compartments (probably divided by hurdling) which was strengthened by three parallel rows of timber uprights and horizontal lacing timbers (*ibid.*: 353). Again, a wedge-shaped ramp gave access to a probable breastwork.

Unsurprisingly, in highland areas of western and northern Britain where stone was plentiful, stone was often (though not exclusively) the sole material used for the construction of ramparts. A particularly fine example of a stone walled hillfort can be seen at Tre'r Ceiri (72) in the Llŷn Peninsula, Gwynedd. This superb site is remarkably well preserved and like the other major hillforts of this area, Garn Boduam and Garn Fadrun, it continued in use into the Roman period. Michael Senior (2005: 16), has suggested that the Romans may have left Tre'r Ceiri alone, either because they deemed it an irrelevance in this isolated part of Britain, or because they were not keen on storming such a strongly defended site.

Also worthy of note are the highly intriguing stone and timber 'vitrified' hillforts found in Scotland. Characteristically, their ramparts consisted of a rubble and earth core that featured dry-stone walling on their inner and outer faces, and which were bonded by horizontal rows of timbers that projected through the outer, and sometimes the inner face of the walls (Cunliffe 2005: 363). It is clear that in many cases, the timber lacings of these ramparts were set of fire causing their core materials to discolour and fuse; hence the term 'vitrified' (*ibid.*) Similar evidence to that found at the vitrified forts has been found at a number of southern British hillforts, with several burnt entrances identified (Avery 1993: 77). It would be strange if further evidence for burning was not found at the many hundreds of Iron Age hillforts in Britain that lie unexcavated, waiting to reveal their prehistoric secrets.

Around the turn of the fourth century BC, after the development of the *Hod Hill* and *Poundbury style* ramparts, which were essentially palisades backed by large sloping banks that contained reinforcing horizontal timbers, the *glacis* or 'dump' rampart became dominant in the south and parts of the south-west (Cunliffe 2005: 364). It appears that these developed as a result of a need to address the problems of rotting timbers that would be difficult to replace, and also the fact that many ramparts were fronted by vertical walls of timber, which would be vulnerable to burning by an enemy force (*ibid.*: 355). Thus ditches were dug deeper and rampart faces sloped back at an angle of 30-45 degrees (*ibid.*), which would create sheer slopes that would not be particularly inviting to any would-be attackers. Notable examples of *glacis* ramparts can be found at Hod Hill and Maiden Castle in Dorset, with the slopes from the bottom of ditches to the top of ramparts measuring 17.4 and 25m respectively (Dyer 1992: 19)

The elaboration of defences

As noted previously, hillforts can comprise of one or more defensive circuits and these different types of fort are termed *univallate* and *multivallate*. It is not exactly clear when multivallation was introduced, but generally hillforts seem to have been given additional ramparts and ditches late in the Iron Age around the end of the first millennium BC (Dyer 1992: 21; Cunliffe 2005: 357). It is likely that this development was largely influenced by military concerns; as James Dyer (*ibid.*: 22-23) says, 'Multivallation was almost certainly introduced to counter spear-thrown firebrands or sling warfare [and] A slinger on top of a high rampart could sling with accuracy about 110m downhill, whilst an attacker slinging uphill had his range severely curtailed'.

Of course, as in any fort, it was the actual entrances that were the weak link in the defensive circuits in Iron Age hillforts and the wooden gates that led into the settlements within them would have been very vulnerable to 'battering and burning' (Avery 1993: 66-67). Britain's Iron Age communities would of course have been aware of this and in several examples hillfort entrances were elaborately designed, in order to make attacking them a hazardous affair. Today, archaeologists hesitate to use such terms as 'military engineers', but it seems beyond dispute that there were skilled and intelligent people among Britain's Iron Age communities who put their minds to the question of defensive architecture.

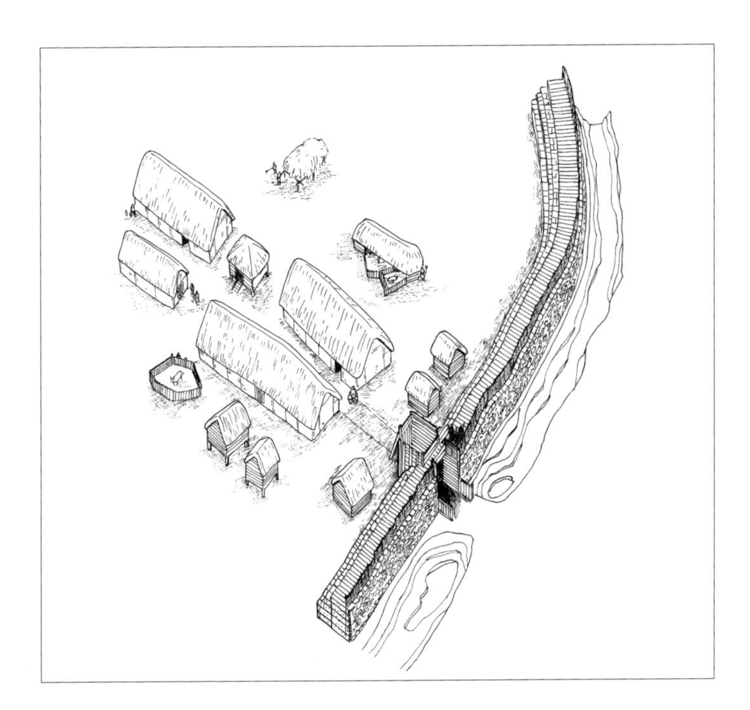

73 Entrance and rampart at first Iron Age hillfort, Crickley Hill (redrawn after Dixon & Borne)

The earliest entrances comprised of a gate or set of gates set at the end of short timber-lined passages, and examples of the former include those found at Hollingbury (Sussex), Ivinghoe Beacon (Buckinghamshire) and the famous Danebury in Hampshire. Examples of the 'dual-portal' entrances have been identified at sites such as St. Catherine's Hill, the Trundle and at Danebury (Cunliffe 2005: 365).

The pattern of postholes found at early entrances implies that many hillfort entrances were also crossed by timber bridges carrying the rampart walk (73) (Dyer 1992: 25). Such bridges would provide not only useful vantage points, but would also provide good positions on which to rain down missiles (and perhaps also boiling liquids) on forces trying to burn or breach gates. It is possible that some entrances featured additional watchtowers, though because of their vulnerability to fire, many scholars have dismissed this notion (*ibid.*).

At some point during the Middle Iron Age in the fifth century BC, many entrances were converted into long corridors (Cunliffe 2005: 374) and a particularly notable example of this can be seen in the final phase (IV) at Torberry in Sussex. Here, the ditch-ends were filled in with walling and a corridor was formed by two huge stone walls that measured 3.0-3.6m wide and that extended into the interior for 27m (74) (*ibid.*: 368). At Danebury, the ramparts were turned outwards to form a 45m long corridor, which had a pair of claw-like

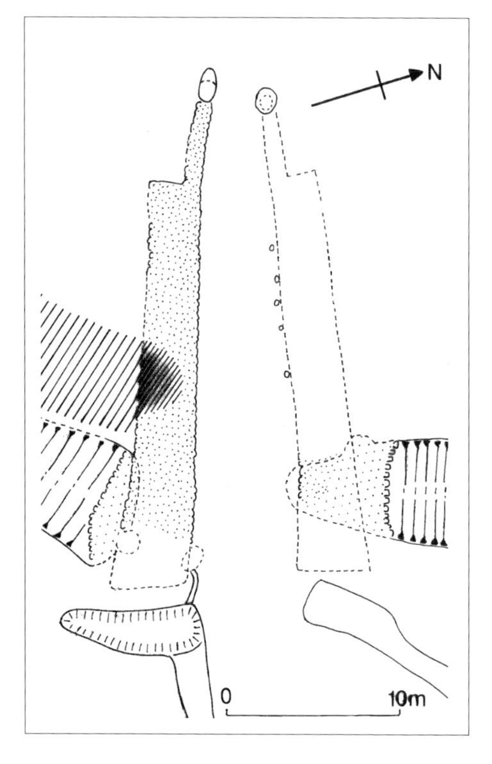

74 Corridor entrance at Torberry hillfort (redrawn after Cunliffe)

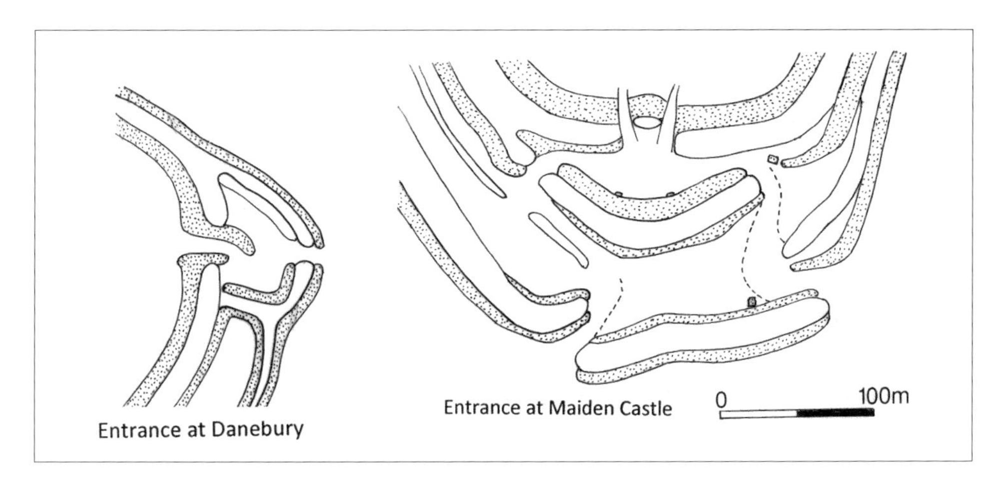

75 Entrances at Danebury and Maiden Castle hillforts (redrawn after Cunliffe)

hornworks protecting an outer gate (*75*) (*ibid.*) Simpler corridors formed from a single outer earthwork which ran parallel to the main rampart can be seen at sites such as Hod Hill and Hambledon Hill in Dorset.

Michael Avery (1986: 224) has plausibly argued that the lengthening of the entrance passages implies that this was a defensive measure that developed out of the need to protect gates from 'fire parties' who would dash up to gates and attempt to burn them during attacks on hillforts. The lengthening of the entrance passage would mean that defenders would be able to stand on the sides of entrance passages and attack fire parties with a barrage of missiles, which in many cases may have simply consisted of a hand-thrown stones (*ibid.*). Thus what would already have been an already risky tactic, even in hillforts with short entrance passages, would have become even more fraught with danger for those attackers who were charged with burning hillfort gates.

In addition to these defensive corridors, elaborate entrances were also built at some southern British hillforts. At the spectacular Maiden Castle hillfort in Dorset, the complex arrangement of earthworks at the entrances (*75*) are still well preserved and give a good idea of how difficult they would make life for any attacking force who attempted to storm the gates. Another fine example of this strengthening of hillfort defences was uncovered during excavations at Crickley Hill. In addition to being the site of the Neolithic 'hillfort' discussed in Chapter Two, two successive Iron Age hillforts were also built here. It appears that the first Iron Age hillfort suffered a similar fate as its Neolithic predecessor, as its gates and houses were burned and its rampart was slighted (Dixon & Borne 1977: 7). Some years after the subsequent abandonment of the site, it was reoccupied and a new and hugely impressive entrance was built. This consisted of a gate defended by two stone bastions and an

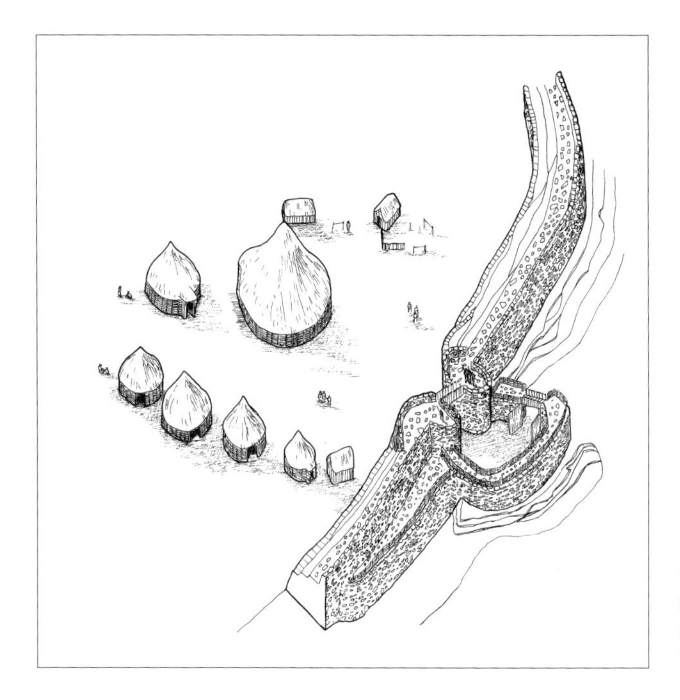

76 Remodelled entrance at second Iron Age hillfort, Crickley Hill (redrawn after Dixon & Borne)

additional stone outwork with a second gate (*ibid.*: 10). This formidable structure would have been reminiscent of the barbicans seen on medieval castles (76).

Many hillforts in North Wales, the Welsh Marches and the southern part of Northamptonshire also had recesses set back immediately behind ramparts, or at the end of long corridor entrances (Cunliffe 2005: 372). These are often – and probably not incorrectly – seen as guard-chambers and it is likely that the men who manned them were responsible for keeping a general watch over the approaches to forts and for shutting gates when an approaching enemy force was sighted (Avery 1993: 87). It is probable that guards kept watch on the day to day traffic entering and leaving hillforts, 'recognising or challenging individuals, refusing entry or exit, reporting suspicious arrivals and departures, and perhaps even exacting market dues in kind' (*ibid.*).

HILLFORT 'HAWKS' VERSUS HILLFORT 'DOVES'

The evidence briefly described above strongly suggests that in many parts of Britain during the Iron Age, communities faced the very real threat of being attacked and responded by building fortifications to defend themselves. It appears that these fortifications were modified through time, in order to try to produce the most effective method of achieving this.

77 Ramparts at Hambledon Hill (Marilyn Peddle)

However, in recent years, the long-held military interpretation of hillforts has been increasingly called into question, with several archaeologists reinterpreting hillforts as symbolic structures and the massive and complex fortifications seen in 'terms of the social isolation and prestige of the dominant elite within' (Armit 2007: 30). It has also been noted by Ian Armit (*ibid.*) that a number of archaeologists have argued that a religious motivation lay behind the construction of hillforts and that the evidence for deliberate burning at hillfort entrances can be explained in terms of ritual rather than military acts. Sue Hamilton and John Manley (2001: 29) for example, have suggested that the ramparts of the South Downs hillforts are unlikely to have been designed with defence in mind and that these hillforts 'might be interpreted as providing focal points for special activities, including storage and cycles of symbolic deposition'. Hillfort entrances which have been cited as having gone up in ritual flames include Moel Hirradug in Flintshire (Bowden & McOmish in Armit 2007: 30) and the final entrance at Danebury (Hill 1996: 108).

The idea that hillforts were used as a means of expressing the status of their occupants is certainly plausible. As Dyer (1992: 26) has rightly said, 'The desire to impress is a very strong human impulse, and an elaborate defence may have been one way of demonstrating social status and wealth'. Therefore, it would certainly be wrong to dismiss 'the symbolic dimensions of hillfort design' (Armit 2007: 33) and it is very likely that the 'overkill' seen in many hillfort defences is related to the desire of their occupants to signal their status or power. However, as Cunliffe (2005: 539) has pointed out in regard to hillforts, 'choice of position, massiveness of defences (77 & *colour plate 25*) and the sheer defensive ingenuity lavished on the entrances is sufficient to suggest that their enclosure works were designed to proclaim the strength of the occupants and to withstand attack'. Ian Armit (2007: 33) has followed a similar line of reasoning; 'it seems perverse to argue that communities which invested so heavily in the construction of massive rampart and ditch systems did so with no thought to defence'. In addition to this evidence, there are the huge caches of slingstones found near entrances at Danebury and Maiden Castle. At the former site, round pebbles were collected from pebble beds that were found between four and six miles from the fort and these were stored in ammunition dumps normally near the ramparts and gates (Cunliffe 2003: 70). Eleven thousand slingstones were stored in a pit near the east gate of the hillfort, and huge numbers of slingstones were also found along the entrance passage, which had either fallen from the outer hornworks, or were hurled there during the defence of the site (*ibid.*). It is interesting to note that the east gate was eventually burnt down in the final century of the first millennium BC. This was probably the result of an attack by a brave enemy force who battered their way through the outer gate and under a deadly rain of missiles, somehow made their way past the defenders on the command post and ramparts flanking the entrance passage to the gate, which they destroyed (*ibid*:. 69). At Maiden Castle, a huge cache of over 22,000 slingstones was found in a pit that lay near a 'sentry-box', which was situated alongside one of the two narrow passages at the east entrance (Wheeler 1943: 47). Many other slingstones were discovered during Mortimer Wheeler's excavations at the site, with a hoard of over 16,000 suggestively discovered near outworks situated just outside the eastern entrance (*ibid.* 49).

Although some have questioned the use of slingstones as weapons (e.g. Hamilton & Manly 2001: 26), it is very improbable that the juxtaposition of stones and ramparts at hillforts is simply coincidental, or that it represents some vague ritual practice. Rather, it is more likely that the elaborate foreworks seen at many hillfort entrances in southern Britain were deliberately designed as fighting platforms for defenders wielding slings and other weapons (Wheeler 1943: 48; Armit 2007: 33). Slingstones have also been found at numerous other Iron Age hillforts in southern Britain (Bryant Finney 2006: 8) and whilst this does not prove that warfare swept over these sites, at the very least, their presence reveals that the threat of attack was a very real possibility.

Turning to the possibility that some of the evidence for burning at hilforts relates to accidental or ritual burning, in regard to the former idea, it seems unlikely that fires would have been built close to wooden gates or ramparts and accidental fires must have been uncommon. It could be than, that many hillfort entrances were indeed deliberately burnt in ritual conflagrations by those responsible for their construction. Interestingly, however, where successive timber and stone-lined entrance passages are found at hillforts, it is the latter that often succeeds the former (Avery 1993: 75). The argument (*ibid.*) that this change from timber to stone was a defensive measure put in place to defend entrances as much as possible from enemy burning is one that carries some weight. Therefore, for this author at least, it seems more likely that much of the evidence for burning points to the attack and deliberate destruction of hillfort defences by hostile forces.

Although they appear to be rare, some British hillforts were given additional protection in the form of closely-set belts of large pointed stones or boulders that were set in the ground outside entrances or other weak points in their defences (Harbison 1971). These outer defences are known as *Chevaux-de-frise* and must have been erected to slow up attackers and make them vulnerable to missiles launched from the walls. *Chevaux-de-frise* are also found in a small number of Irish Iron Age forts, although here the stones often run all the way around walls, as most spectacularly seen at the superb cliff-top fort of Dun Aenghus in County Galway. Generally, the stones of the *Chevaux-de-Frise* reach about 0.5m in height, although at Pen-y-Gaer in the Conwy Valley some stones reach 1m and at Dun Aenghus there are stones nearly 2m in height (*ibid.*: 209). The existence of a wooden *Chevaux-de-frise* is strongly suspected at South Barrule hillfort on the Isle of Man. It is not improbable that many other wooden examples once stood outside British hillforts, although they have been lost because of burning or decay (*ibid.*: 212), or not identified during excavations.

Of course some hillforts may well have functioned as such things as sacred precincts, political centres, elite residences or even as the sites of fairs but this does not preclude them from being deliberately fortified sites (Armit 2007: 36). All in all, the evidence suggests that hillforts were places that were deliberately designed to deter potential enemies from attacking the Iron Age communities who resided or gathered within them and that on a number of occasions these attacks actually took place.

BROCHS: DEFENDED STRONGHOLDS OF THE SCOTTISH IRON AGE?

No matter how brief, any discussion of fortifications in Iron Age Britain should not pass over the unique circular stone towers known as brochs that are commonly found in northern and western Scotland. These intriguing and impressive structures appear to have originated around 400 BC, although the evidence suggests that their main period of use was from the second century BC to the second century AD (Cunliffe 2005: 335).

Francis Pryor (2003: 391-392) has provided a succinct and useful summary of the architectural features of brochs: 'Their main characteristic was a hugely strong hollow dry-stone wall, through which ran stairs. The interior of the tower was most probably roofed, and the wooden supporters for an upper floor, or floors, were lodged on an internal ledge known as a scarcement. Each floor was reached by a narrow entranceway through the inner 'skin' of the wall. The cavity within the wall contained small cell-like compartments and galleries at each floor level, and there was always a substantial 'guard-chamber' at ground level close to the front door. This was the sole entrance, and it led into a corridor which passed straight through the wall into the interior'.

As Ian Armit (2003: 112) has remarked, it is hardly surprising that broch towers have traditionally been identified as defensive strongholds, as on first sight they resemble later castles and tower-houses, which marked the fortified seats of clan chiefs. However, there are valid arguments that can be raised against this traditional view, and as Francis Pryor (2003: 393) has argued, attackers could easily climb broch walls, roofs were unprotected from such attacks, doors could easily be rammed, and perhaps most tellingly, there were no windows or slots to monitor hostile activity outside the broch.

Therefore, it is perhaps more probable that the actual defence of brochs had more to do with their location and their associated substantial stone ramparts and ditches, rather than the imposing towers themselves (Armit 2003: 113). Notable examples of such defensive works can be seen at the Broch 'villages' of Gurness (78) and Howe on Orkney and at the impressive Clickhimin in Shetland. The latter site was built on an islet, lending further support to the

78 The Broch of Gurness (Sigurd Towrie)

79 Clickhimin broch (redrawn after Sorrel)

idea that some brochs were built with defence in mind (79). Nevertheless, even brochs that stood alone without additional fortifications still possessed defensive qualities. As Pryor (2003: 395) has noted, a broch would present an imposing sight to any would-be attackers and would have provided short-term defence from raiders.

Ian Armit has put forward the interesting idea that like the heavily defended hillforts of southern Britain, the brochs of Scotland could have some connection with the Roman demand for slave labour which had destabilised the native economy in many areas of France (or Gaul, as it was known in first century BC). Just as the Gaulish elites had swapped slaves for luxury goods such as wine, so too may their Iron Age counterparts in Britain have been involved in similar prestige-enhancing trade, with organised and speculative raids for slaves taking place the length and breadth of Britain, perhaps as early as the mid first century BC (*ibid.*). As Barry Cunliffe (2004: 100) has remarked, 'it is *possible* that the proximity of the Roman market encouraged native communities to generate slaves as a new cash crop'. Although we have no way of knowing the extent of the slave trade in Late Iron Britain, there can be little doubt that one existed; not only do Classical authors such as Strabo mention that they were one of Britain's exports to the continent during the Iron Age, but there have also been actual finds of slave chains at Llyn Cerrig Bach, Anglesey and Bigbury hillfort in Kent (*ibid.*). Miranda Aldhouse Green (2004: 337), has pointed out that slave or 'gang-chains' such as those found at Llyn Cerrig Bach would not just have been a useful means of restraining war captives, but they could also be used in the symbolic humiliation and degradation of the enemy who ultimately were dehumanised by their imprisonment in such chains. The slave chain from Llyn Cerrig Bach was part of a remarkable collection of Iron Age metalwork, much of it militaristic in nature, and it is quite possible that this metalwork represents war booty offered to deities (Cunliffe 2004: 106).

Of course, it would not just have been people that hillforts protected and various economic resources were probably a major driving force behind warfare. It has been suggested (Cunliffe 2003: 76) that cattle were probably often the target of Iron Age raiders, as depicted in the early Irish tale, *Táin Bó Cúailnge* or 'the cattle raid of Cooley' which began as a cattle raid, but spiralled into more serious warfare. Although the *Táin* dates to the seventh century AD, it nevertheless provides us with a good picture of an unstable, warfare-prone Iron Age society largely unaffected by Rome and a similar situation may have existed in the Late British Iron Age around 100 BC (*ibid.*). Archaeological evidence also indicates that some hillforts acted as redistribution centres where other goods such as grain and wool were stockpiled and where rarer goods such as salt, iron and shale were brought to be worked into valuable items (Cunliffe 2005: 391). Such 'wealth' may well have attracted the unwelcome attention of 'foreign' raiders who were prepared to risk their lives in attacks on hillforts as they sought to capture the various goods that lay behind the ramparts of some sites

CRANNOGS

Although they lack the formidable presence of brochs, the 'crannnogs' or island roundhouse dwellings that were also common throughout Scotland in the Iron Age were probably defensive in nature. These sites consist of foundations of timber, peat, brushwood or stones dumped on loch or river beds, or added to existing natural islands on which the roundhouses stood (Barber & Crone 1993: 520). Whilst crannogs are a characteristic feature of the Iron Age, it is evident that they have a very long history. The Neolithic island dwelling at Loch Olabhat that was mentioned in Chapter Two represents the earliest known example (Henderson 1998: 229) and it appears that people were still building and living on crannogs in Scotland and Ireland in the seventeenth century (*ibid.*).

WEAPONS, WARRIORS AND THE VICTIMS OF WAR

It seems appropriate to begin this section with a brief discussion of slingstones, which as we have seen above, are well represented at Iron Age hillforts and appear to have been an important weapon of war. In fact, as Jon Bryant Finney (2006: 6) points out, in the Middle Iron Age (*c*.400-150 BC) at least, aside from pottery, slingstones must be one of the most common archaeological finds from this period. Slingstones were made from four different materials, with natural water-worn pebbles being the most common type used, although baked and unbaked clay ovoids were also used alongside occasional examples carved from chalk (*ibid*.).

The actual slings themselves would have been made from materials such as wool, cotton, leather, or even plant fibres, grasses and rushes (*ibid*.: 67). Therefore it is hardly surprising that we have no examples from Iron Age Britain and it seems highly unlikely that a sling from this time will ever be found. Barry Cunliffe (2005: 534) tells us that 'The sling could be used in two modes, swung in a vertical plane it could send volleys of stones into the air to rain down on attackers at some distance, while used in a horizontal swing stones could be sent at considerable velocity into a body of troops accurately at head height with devastating effect'. As has been noted (Cunliffe 2003: 70-71) modern ethnographic parallels also suggest that Iron Age slings may have been used to drive off wild animals preying on flocks or in the herding of animals and it is probably in these contexts that the use of the sling first arose.

Daggers and swords

Some of the finest artefacts left behind by Iron Age smiths as a testimony to their skill and artistry are the superbly made decorated daggers and swords. The daggers date to the Early Iron Age *c*.600-400 BC and have mainly come form various stretches of the Thames (Jope 1961). The style of the earliest daggers shows that British smiths were influenced by the 'Hallstatt D' daggers produced in Germany in the early sixth century BC, some of which probably arrived in Britain as 'diplomatic gifts' from Iron Age leaders on the continent (Cunliffe 2005: 462). A possible example of one such gift may be provided by the imported dagger found in the Thames at Mortlake (*80*) (*ibid*.). Although similar in style to the continental daggers, the British examples feature twin suspension loops on their back plates in contrast to the single loops found on the former (Jope 1961: 307). Whether these daggers were ever actually used in combat is certainly debatable, but it remains a possibility nevertheless and they would undoubtedly have been lethal weapons in fighting at close quarters. The later La Tène daggers also reveal contacts with the continent and the origin of

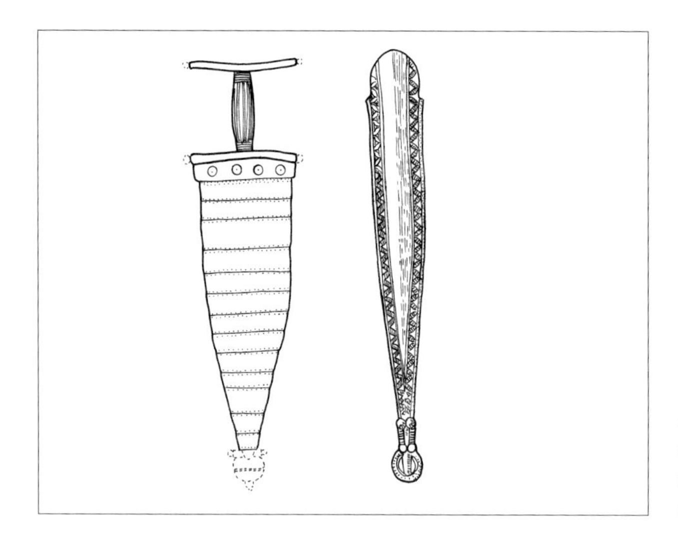

80 Imported Hallsatt dagger and La Tène scabbard of the Early Iron Age (redrawn after Jope)

these daggers can probably be found in eastern France (*ibid.*: 312). They differ somewhat from their earlier Hallstatt counterparts in that they are more slender and feature distinctive chapes on the ends of the scabbards (*80*). Again, they would have been the preserve of those who sat at the head of Iron Age society and it is not impossible that on occasion they were also used in warfare.

While there is some uncertainty as to whether daggers were used in battle, there can be little doubt that swords were an important weapon of war during the Iron Age. The earliest swords were close copies of the bronze Gündlingen type, which were probably introduced into Britain from the continent during the Bronze Age/Iron Age transition in the first half of the eighth century BC. (Cunliffe 2005: 449). These long swords were sheathed in scabbards of wood or leather which terminated in distinctive winged metal chapes, which may have been deliberately designed so that a mounted warrior could keep the end of his sword sheath steady with his left foot, while drawing the weapon within it with his right hand (*ibid.*). Of course, this theory assumes that mounted warriors equipped with swords carried them at their sides, with weapons attached to belts of some kind. However, there is strong evidence from the archaeological record which indicates that at least in some areas of Britain, warriors actually wore swords on their backs. For example, in a warrior burial found between Burton Fleming and Rudston in Yorkshire, a sword lay underneath the torso of the male skeleton which lay extended in the grave (Stead 1988: 23). In addition to this discovery, there is also a collection of fascinating carved chalk figurines, also found in Yorkshire, some of which depict warriors with back-mounted swords; it has been suggested that these figurines may 'depict one distinct individual – a warrior-god, mythical figure, or ancestor' (*81*) (*ibid.*: 25).

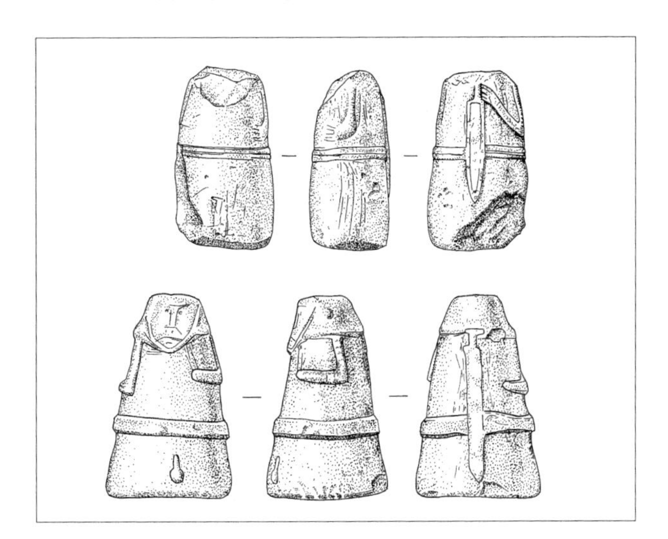

81 Carved chalk warrior figurines from Yorkshire (redrawn after Stead)

By the late fifth or early fourth century BC, long iron swords of the 'La Tène' type had supplanted the Gündlingen swords and the short swords and daggers that had been in vogue in the earlier centuries of the Iron Age (Cunliffe 2005: 533). The early La Tène I swords had blades better suited to thrusting and slashing, while the La Tène III swords had blades more suitable for slashing. This could indicate that fighting on horseback became more important in the latter centuries of the Iron Age (*ibid.*). Many La Tène swords were beautifully made and decorated, with bronze often used for scabbards, although iron was also used and it is not unlikely that on occasion, wood and leather scabbards were favoured over the more 'expensive' metal versions (Piggot 1950: 2). One of the finest examples of a La Tène sword is the example found at Kirkburn in Yorkshire, which 'is a masterpiece of British craftsmanship probably dating to the third century BC' (*colour plate 26*) (Cunliffe 2004b: 43).

Spears

Another important weapon during the Iron Age was undoubtedly the spear and as in the Bronze Age, a wide variety of spearheads were made during this time (*82*). Again, they can be broadly divided into lighter 'javelins' for throwing and heavier 'lances' for thrusting. The former would have been particularly effective when thrown in volleys against the enemy, as a spear could not only wound and kill enemy combatants, but also lodge in shields and greatly impede movement on the battlefield, with warriors perhaps sometimes being forced to throw away their shields (Cunliffe 2005: 534). Of course, as with Bronze Age spears, it is improbable that such rigid distinctions were always made when lances and

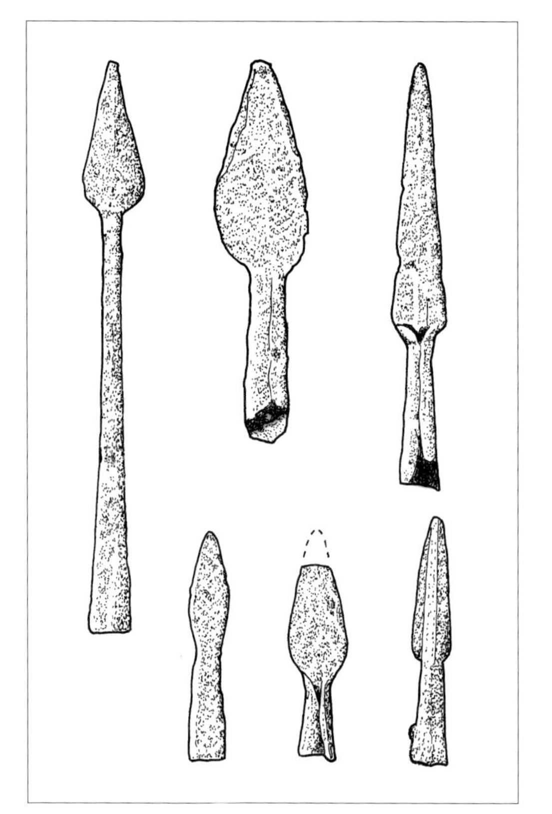

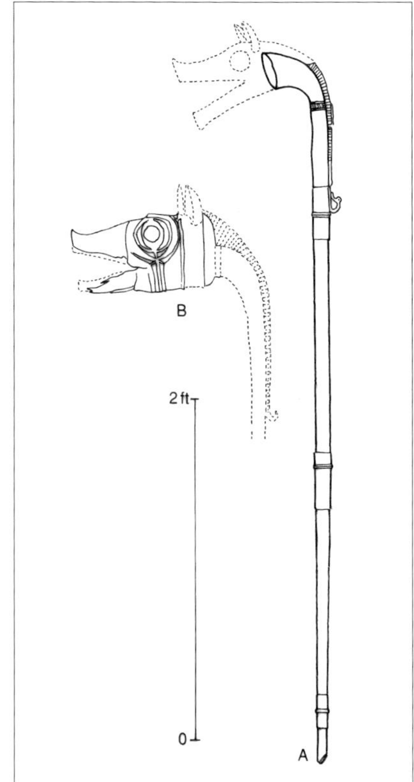

Above Left: *82* Iron Age spears and javelins (redrawn after Cunliffe)

Above Right: *83* Carnyxes from Tattershall and Deskford (redrawn after Piggot)

javelins were used in battle and each may sometimes have performed the other's function if the need arose. However, in Britain, lances may primarily have been a cavalry weapon, as suggested by the depictions of 'lancers' seen on some British Iron Age coins (*ibid.*).

In addition to the actual weapons that brutally took lives in Celtic Warfare, an important psychological weapon in this Warfare was noise (Cunliffe 2005: 535). In his account of the famous Battle of Telamon fought in Italy in 225 BC, Polybius paints a frightening picture of the Celts in Battle (*ibid.*):

> [the Romans] were terrified by the fine order of the Celtic host, and the dreadful din, for there were innumerable hornblowers and trumpeters and, as the whole army were shouting their war-cries at the same time, there was such a tumult of sound that it seemed that not only the trumpeters and the soldiers but all the country round had got a voice caught up in the cry.

Two rare examples of Celtic war-trumpets or *carnyxes* have been discovered in the River Witham at Tattershall Bridge, Lincolnshire and at Lieceston farm, in the parish of Deskford, Banffshire (*83*) (Piggott 1959).

DEFENSIVE EQUIPMENT

As in the later Bronze Age, two different types of shield were made in the Iron Age; those used in actual combat and those that were made primarily as ceremonial items to be used in rituals and displays of status. The former type of shield is sub-rectangular or oval in shape, made from leather or wood (or a combination of the two) and featured plain metal fittings such as bronze edge bindings (Cunliffe 2005: 535). The latter feature numerous sheet bronze facings and were superbly decorated with the flowing repoussé and engraved motifs which are so characteristic of Celtic La Tène art. The all-metal shield found in the Thames at Chertsey in Surrey featured 9 separate bronze parts, while the

84 Statue of a Celtic warrior from Vachères (author)

stunning shield facing discovered near Battersea comprised of over 90 (James & Rigby 1997: 41) (*colour plate 27*).

Two rare sheet bronze helmets dating to the Iron Age have been found in Britain, with one of unknown provenance in the Meyrick collection and a superb horned example found in the Thames at Waterloo Bridge (Cunliffe 2005: 535). It is more likely however, that as with the metal shields, these helmets were primarily made for display and that well-padded helmets made from thick leather were the more normal head-gear worn in battles as they would have been far more effective at deflecting blows (*ibid.*).

It is evident that some warriors also wore chain mail shirts for protection in the Iron Age. Examples of such mail have been found at the Kirkburn 'chariot' burial mentioned below, in elite burials at Lexden, Folly Lane and Baldon, and chain mail fragments have been discovered at a number of other sites (*ibid.*). Some idea of the appearance of a mail-clad Celtic warrior is provided by a fine late first century BC statue found at Vachères in France that depicts a Gaulish warrior wearing a mail shirt and cloak, and accompanied by a shield and sword which hangs from his belt on the right-hand side (*84*).

CHARIOT WARFARE

Although the use of the war chariot appears to have died out among the Celtic tribes of continental Europe after the Battle of Telamon in 222 BC, a famous passage written by Julius Caesar reveals that some 200 years later they were still in use in Britain (Berresford Ellis 2003: 68). Caesar tells us (with obvious respect for the enemy that he was facing) that:

> In chariot fighting the Britons begin by driving all over the field hurling javelins, and generally the terror inspired by the horses and noise of the wheels is sufficient to throw their opponents' ranks into disorder. Then, after making their way between the squadrons of their own cavalry, they jump down from the chariots and engage on foot. In the meantime their charioteers retire a short distance from the battle and place the chariots in such a position that their masters if hard pressed by numbers, have an easy means of retreat to their own lines. Thus they combine the mobility of cavalry with the staying power of infantry; and by daily training and practice they attain such proficiency that, even on a steep incline, they are able to control the horses at full gallop, and to check and turn them in a moment. They can run along the chariot pole, stand on the yoke, and get back into the chariot as quick as lighting.

There is little reason to doubt the authenticity of Caesar's account which provides us with a vivid snapshot of Celtic warfare. Of course, the Iron Age forces mentioned in it were fighting the Roman legionaries, but it is unlikely that war chariots were not sometimes used in earlier pitched battles that involved only British Iron Age tribes. As in the Bronze Age (and perhaps earlier) single combat between individual champions would also have been a feature of Iron Age warfare and such encounters would have been a prelude to larger and serious clashes, though in some cases they may have been enough to settle the grievances of opposing forces (Cunliffe 2003: 76). Those who were involved in such duels would have belonged to the warrior caste that was clearly an important element in Iron Age society and warrior burials are not unknown in Britain. One of the best known comes from the Iron Age cemetery at Mill Hill near Deal in Kent, where a warrior was buried with a sword, a large leather-covered wooden shield, and a unique 'crown' made from bronze strips (*85*) (James & Rigby 1997: 24).

Possible examples of Iron Age war chariots have been found in a limited number of burials in the distinctive cemeteries of the Middle Iron Age 'Arras Culture' of eastern Yorkshire, which might have developed as a result of the arrival of newcomers from northern France or Belgium at some point in the fifth century BC (Cunliffe 2004: 43). Notable finds in this regard are two burials discovered at Wetwang Slack in the summer of 1984, where a young male and young adult (probably male) were laid on the bodies of box-like vehicles; above these were placed the two wheels of the vehicles (Dent 1985). The young man was accompanied in death by grave goods that included a sword, seven spearheads, iron coverings from a wooden shield, probable harness-fittings and the forequarters of a pig. Likewise, his counterpart was buried with a sword, a probable iron reinforcement for a wooden shield boss and horse harness. Not far away from these two graves another vehicle burial was discovered at Kirkburn in 1987 (Cunliffe 2005: 549). Here, the deceased had been buried with a coat or iron mail, two sets of pig bones and the wheels and yoke of the vehicle (*ibid.*)

The evidence found in these burials strongly indicates that we are looking at the burials of high status warriors, but there is however, some question as to whether the vehicles found with these warriors are actually war chariots. As Barry Cunliffe (2004: 98) has pointed out, the evidence is against this idea; only four of the 15 or so vehicle burials that have been excavated were accompanied by weapons, the burial rite included females buried without weapons and the box-shaped structure of the vehicles suggests a heavier cart rather than an actual light-framed war chariot. Nevertheless, it is quite possible that these vehicles had a dual purpose and that their chassis could have been adapted for the purposes of warfare, and turned into chariots if required (*ibid.*: 99).

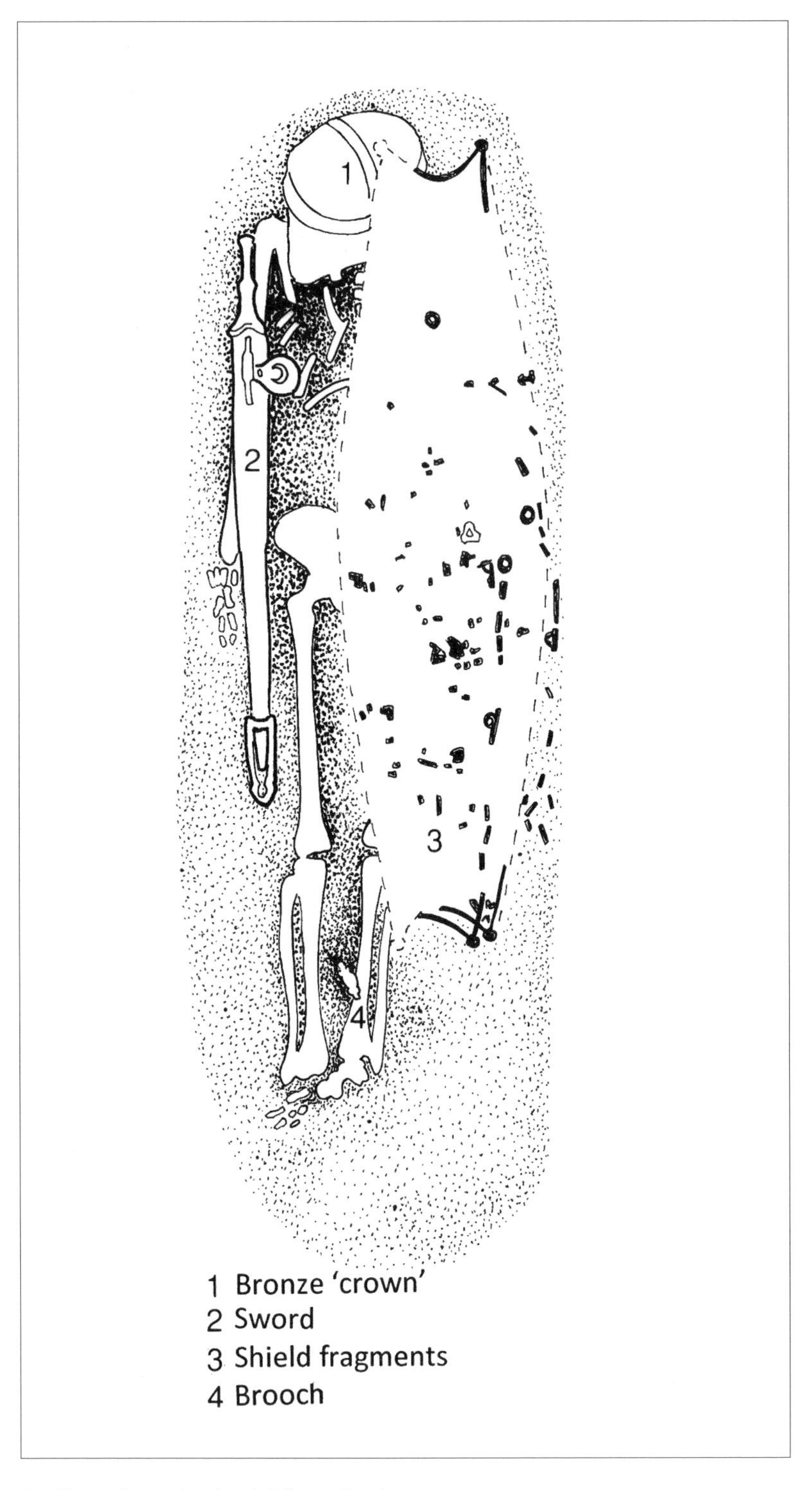

85 Plan of warrior burial from Deal

86 Danebury pit burials (Danebury Trust)

SKELETAL EVIDENCE

It has been said that Iron Age battles 'were horrific affairs resulting in death and mutilation' (Cunliffe 2003: 77) and human remains from Iron Age sites bear this statement out. A good place to start our brief look at these remains would be the famous Danebury hillfort, which has yielded a number of likely battle casualties.

Among the huge mass of archaeological evidence to come from this fascinating and superb site were many human remains (representing at least 100 people) that were found buried in grain silos and other contexts within the hillfort (*86*). These burials may well represent propitiatory offerings to chthonic gods to ensure the fertility of crops (*ibid.*: 147). The remains consisted of 25 complete bodies along with many partial ones, and separate skulls and disarticulated limbs were also interred during the Danebury 'pit-tradition' (Aldhouse Green 2002: 146). A number of people bore terrible injuries that had obviously been caused by weapons (Cunliffe 2003: 77). One man had been viciously assaulted, and had had a spear thrust into his forehead in addition to suffering two other severe blows to the head. Another male aged around 30 was probably struck in the centre of his forehead with a sword, though this blow did not immediately kill him and he survived for some time afterwards. Another individual received a brutal blow above the right eye socket and this probably destroyed his eye, while a human femur displaying sword cuts was also found (*87*).

Interestingly, the human remains found at Danebury show a distinct bias towards men in their twenties (Aldhouse Green 2002: 146), which lends some weight to the idea that at least some may have been warriors who had lost their lives in warfare. However, women and children were also represented among the skeletal material and it has been suggested (Craig *et. al.* 2005: 170) that two decapitated juvenile skulls found in Pit 2509 may have been displayed before burial. Several other isolated human skulls were found in 15 pits and it is quite possible that they represent evidence of the Celtic practice of headhunting (Cunliffe 2003: 151) recorded by many Classical authors, but perhaps most famously by Diodorus Siculus, who tells us:

> They cut off the heads of enemies slain in battle and attach them to the necks of their horses ... The bloodstained spoils they hand over to their attendants to carry off as booty ... and they nail up these first fruits upon their houses ... They embalm in cedar oil the heads of the most distinguished enemies, and preserve them carefully in a chest and display them with pride to strangers.

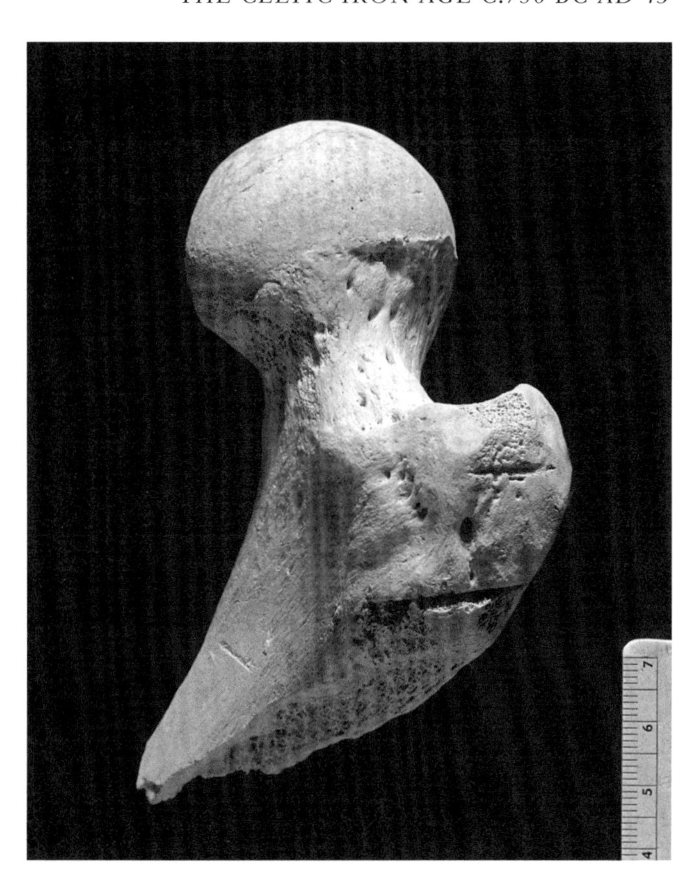

87 Hip bone from Danebury showing sword cuts (Danebury Trust)

Exactly why the Celts practiced headhunting is unclear, but it is likely that 'The head was regarded as the seat of a man's power and ownership of a head meant control over that power' (Cunliffe 2003: 154). Celtic warriors may also have felt that they were controlling the vengeful spirit of the deceased by taking it from the body (James in Berresford Ellis 2003: 167).

Of course, not all of the human remains found at Danebury can be interpreted as representing casualties of war, but the evidence does strongly suggest that several people lost their lives as a result of warfare, although whether they were warriors or civilians who were members of the Danebury community is unclear. Miranda Aldhouse Green (2002: 146) has made the interesting suggestion that some of the bodies may have been 'honourable battle-dead', as they were buried behind the ramparts. It has also been argued (Craig *et. al.* 2005: 176) 'that some of the mortuary remains from Danebury (and perhaps other Iron Age sites in Britain, such as those from South Cadbury) provide ample evidence for mass killing and mass burial, peri-mortem mutilation and dismemberment, and display of bodies and body parts that is in keeping with more recent instances of revenge warfare'. It is possible then, that at Danebury, some of the human remains may represent 'bodily talismans of vanquished foes rather than revered ancestors' (*ibid.*). In this respect, Keith Otterbein (2000: 439) has pointed out that many societies kill captured enemies, including women and children. Interestingly, some of the complete bodies at Danebury had tightly-flexed limbs, indicating perhaps that these people had been bound like prisoners before burial (Aldhouse Green 2002: 146). It could be that these people represent war captives dedicated to an Iron Age deity or deities.

Another example of the ritual sacrifice of a war-captive to supernatural forces can perhaps be seen with the famous bog body of 'Lindow Man' found at Lindow Moss in Cheshire (*colour plate 28*). Examination of the remarkably preserved body revealed that Lindow Man was about 25 when he had suffered a terrible death; he was struck hard on the head twice, strangled and had his throat cut before being thrown naked (apart from a fox-fur armband) into a marsh (*ibid.*: 124). Intriguingly, he had traces of copper compounds on his skin and these could possibly represent the remains of blue body paint such as that Caesar reported seeing on the bodies of naked British warriors when he landed in 55 BC (James & Rigby 1997: 24). Therefore, it is quite possible that Lindow Man was a foreign hostage or captive taken in war who was 'brought low and ritually slaughtered by an alien community' (Aldhouse Green 2002: 124).

Further evidence pointing towards headhunting in Iron Age Britain has been discovered at other sites. For example, at the hillfort of Breedon Hill, Gloucestershire, the remains of human skulls found with burnt material at the hillfort's entrance could represent human heads that had hung above the gateway, but which subsequently fell down when it was set on fire (Hencken in Craig *et. al.* 2005: 172). Similar evidence has come from South Cadbury in Somerset, where burnt human skull fragments were interpreted as the decapitated heads of the vanquished that had been displayed near the destroyed gate and later burnt by those who attacked and overran the hillfort (Woodward & Hill in Craig *et. al.*: *ibid*). Human heads may also have been hung above the entrance to Stanwick hillfort in Yorkshire and an Iron Age skull fragment found at a broch at Hillhead in Scotland has three holes drilled in it, indicating the suspension of a human head (Aldhouse Green 2002: 104).

At Eiskerton in Lincolnshire, an adult male skull displaying a sword cut was found in a ritual deposit dating to the fourth/fifth centuries BC (*ibid.*) and likewise, at Acklam in east Yorkshire, a male skeleton found with a sword in a grave appears to display severe sword wounds on the back of the skull (Dent 1983: 124). A number of other Iron Age burials in Britain have produced other examples of people wounded or killed by swords (*ibid.*: 124-125).

Excavations at Iron Age cemeteries in East Yorkshire have revealed a remarkable burial rite which consisted of driving spears into the dead as they lay in their graves (Stead 1991). However, it is likely that the spears found in two graves at Rudston had actually killed the individuals who had been interred in them. In grave R152, one of the deceased had been speared from behind, with the spear probably penetrating the heart. Likewise, in grave R94, a spear had been thrust into the back of the deceased, entering the body just to the right of the spinal column. It is also possible that the person buried in grave R140 had been killed by a spear that had been thrust with some force through the right pubic bone (*ibid.*: 136-137). Finally, another likely example of an individual killed by a spear is the female aged 25-30 years found at Wetwang Slack, who appears to have been killed by a spear-thrust to the stomach (Dent 1983: 125).

Pennal

CHAPTER 6

CONCLUSION

It would be foolish to argue from the material briefly examined in this book that prehistoric Britain was a bloody and brutal place in which warfare was a constant feature of life for its various inhabitants. Although there is a good deal of archaeological evidence pointing towards warfare, we must be careful of seeing it everywhere we look simply because the archaeological pendulum has swung firmly towards the idea that the prehistoric world was a dangerous and violent one (Parker Pearson 2005: 21). We should however, be equally careful of following the pendulum back the other way, with fortified sites, weaponry and skeletal evidence interpreted in terms of 'status' and 'ritual', rather than as indicators of actual warfare.

It should be remembered that there is plentiful evidence from the ethnographic record which reveals that warfare in non-state societies was a frequent and often brutal aspect of life that significantly impacted on communities (see Keeley 1996: 83-97; Gat 2006: 129-131). In fact, the numbers of people who lost their lives in 'primitive wars' was considerably higher than the numbers killed as a result of 'civilised wars'; war-related mortality in non-state societies averages between 20-30 per cent, which stands in marked contrast to state societies where it is below 5 per cent (Helbling 2006: 114).

Some of the skeletal evidence in this book, may of course, represent acts of violence rather than warfare and we must consider such possibilities as murder, violent brawls, the execution of criminals, the sacrifice of community members and even accidents. Nevertheless, whilst this seems likely in regard to some of the examples included in this book, all possible skeletal evidence for warfare should be considered. In any case, many of the individuals displaying lethal trauma probably represent people who were in one way or another involved in the sphere of warfare, whether they were warriors killed in combat, non-combatants killed in attacks, or prisoners of war who suffered the horror of ritual murder. As Jonas Christensen (2004: 135) has said, 'perimortem trauma on skeletons is the most direct evidence of the consequences of war'. Furthermore, the anthropological record has shown that in non-state societies, acts of interpersonal violence between individuals often escalated into larger-scale warfare involving whole villages or other large groups (Maschner & Reedy-Maschner 1998: 20).

There can also be little doubt that many victims of warfare in prehistoric Britain have gone unrecognised but these 'invisible' victims of war are often ignored, however. Numerous people must have lost lives as a result of lethal injuries to major organs or arteries which have left no traces on skeletons. Also, without the benefits of modern medicine, even superficial war wounds and injuries would pose a serious threat to those who had received them, and septicaemia must have taken numerous lives. The ethnographic record suggests that some cases of blood poisoning may have been brought about as a result of being struck by poisoned weapons. For example, in North America, warriors commonly tipped their arrowheads with snake venom, and poisonous plants such as hemlock were also used (Keeley 1996: 52). The Nevada Shoshoneans employed more 'creative' methods when it came to poisoning their war arrows and used blood from the heart of a mountain sheep, which had been poured into a section of its intestine that had been buried in the ground and left to rot (*ibid.*). The Mae Enga of Highland New Guinea attached hollow cassowary claws over the ends of their arrows and spears, which would remain in the wound and cause infection, while the Dani of the same region daubed mud or grease on their war arrows to the same end (*ibid.*).

Another thing that should be taken into account when considering the unrecognised victims of warfare in prehistoric Britain is that many may have gone unrecorded because antiquarian 'excavations' often ignored skeletal evidence in favour of the artefacts that were found accompanying prehistoric burials. However, to say that the antiquarian diggers had no concern for the prehistoric people that they discovered would be doing them a disservice. For example, antiquarians such as Colt Hoare, William Cunnington and John Thurnam recorded many instances of weapons injuries in the Neolithic tombs and Bronze Age barrows that they dug into (Thorpe 2005: 146). Although their accounts are often discredited as the result of over-active imaginations, as Nick Thorpe (*ibid.*) has pointed out, without re-examination of these skeletal remains, they should not be simply dismissed.

Staying with those victims of prehistoric warfare in Britain who may have gone unrecognised, James Whitley (2002: 122) has argued that all too often, human remains found on prehistoric religious sites are seen as belonging to the world of the 'ancestors', when in fact some may represent 'social outcasts' or the 'unquiet dead', rather than the revered dead. It is quite possible that some of these unquiet dead were war captives who were ritually murdered in honour of prehistoric deities, or even people who were killed in attacks on these sites.

TOWARDS AN UNDERSTANDING OF PREHISTORIC WARFARE IN BRITAIN

We can of course, never uncover the whole truth behind the form that armed conflict took in prehistoric Britain, nor the reasons why its communities fought each other in these conflicts. Nevertheless, some suggestions have been put forward in this regard and as we have seen, it appears likely that warfare often arose over resources of various kinds. However, warfare would not have been solely driven by 'economic' concerns and the ethnographic record can not only provide us with further possible insights as to other causes of warfare in prehistoric Britain, but also indicate the forms it took.

Anthropological accounts have shown that the most common and deadly form of warfare in non-state societies around the world was the surprise raid or attack, which often took place at night or just before dawn (Helbling 2006; Gat 2006; Keeley 1996). Raids could be carried out by small groups of warriors with a single, or limited number of individuals targeted, although they could also be large-scale affairs featuring whole clans or tribes who attacked and indiscriminately massacred the inhabitants of enemy settlements. Although the smaller scale raid was often favoured and only a few people were killed at a time, their effects should not be underestimated; women and children could lose their lives and the frequency of raids meant that fatalities rapidly accumulated. It could be argued that raiding does not amount to warfare *per se*, and that it should not be placed in the same bracket as war. However, as John Kennedy (71: 53) has rightly asked: 'How different is a commando raid or a "search and destroy" mission in Vietnam from a raiding party of the Yanamomo, the Dane, the Nuer or the Jivaro? Maybe you too have noticed how the essentially age-old "primitive" tactics of war have suddenly been made respectable and modern, when reinvented by such military "geniuses" as Mao Tse Tung and Che Guevara'.

In addition to the more deadly warfare waged by non-state societies, it is evident that mutually arranged ritual, or 'formal' battles took place. As Lawrence Keeley (1996: 59) has pointed out, anthropologists have often focused on these battles rather than other types of combat, which led to the traditional view of 'primitive' warfare as being little more than a dangerous sport. A brief look at a formal battle fought by clans of the famously warlike Mae Enga of highland of New Guinea (Meggit 1977: 17-21) will provide us with some idea of the ceremonial nature of such encounters.

The battle opened with the hurling of insults, chants and stylised aggressive displays by either side. After this warriors of renown met in the middle of the battlefield and proceeded to fight with shields and spears. However, fatalities in such encounters were low because opponents were often equally matched and aimed to wound, rather than to kill. After the 'duel' was over, gifts were exchanged and these could include plumes of feathers, shells and stone axes.

The next stage in the battle consisted of opposing skirmish lines letting loose their bows, as they ran hither and thither on the battlefield, with each force aiming to outflank each other, in order to achieve a deadly crossfire. However, a stalemate normally ensued because both sides used the same tactics and the rules of battle did not allow combatants to leave the field. This constrained skirmishing normally continued until dusk, when, by mutual consent, fight leaders would stop the conflict.

Generally therefore, the Mae Enga formal battles normally ended in a 'draw', with a limited loss of life on both sides. Nevertheless, while 'this is true enough … we in the comfort of our armchairs, should not forget that men died in great fights, painfully' (*ibid.*: 21).

Another example of formal-style warfare is provided by Saukamapee, a Cree Indian who lived among the Piegan tribe of the Blackfoot, on the North American Great Plains in the eighteenth century. Saukamapee says:

> After some singing and dancing, they sat down on the ground, and placed their shields before them, which covered them. We did the same … their bows were not so long as ours … and their arrows went a long way and whizzed about us as balls do from guns. Our iron-headed arrows did not go through their shields, but stuck in them. On both sides, several were wounded, but none lay on the ground; and night put an end to battle without a scalp being taken on either side and in those days such was the result, unless one party was more numerous than the other. The great mischief of war then, was as now, by attacking and destroying small camps of 10 to 30 tents, which are obliged to separate for hunting (Tyrell in Bamforth1994: 99).

Therefore, as Douglas Bamforth (*ibid.*) notes, Saukamapee reveals the existence of two distinct types of warfare on the Great Plains; one that involved somewhat ceremonial battles with massed shield lines and few casualties, and a more deadly one that consisted of the systematic destruction of small Indian communities.

Although we lack space here to go into the varied and complex causes of warfare in non-state societies, it is evident that there were common reasons why wars broke out among 'primitive' peoples. Aszar Gat (2006: 92) has stated that 'Revenge has probably been the most regular and prominent cause of fighting cited in anthropological accounts of pre-state societies. Violence was activated to avenge injuries to honour, property, women, and kin'. Of course in many cases, those who set out to redress such wrongs did not desire to start wars, but only wished to kill those responsible for the original offence. However, things were often not that simple and such revenge killings could quite easily escalate into more serious episodes of armed conflict.

Less prosaic concerns could also promote warfare between non-state peoples and lead avenging warriors to seek out wrongdoers. For example, a band of Arunta hunter-gatherers in the central Australian desert sent out an *atinga* (avenging party) of warriors to avenge several recent and unexplainable deaths in their group after their 'shaman' had determined that they had been caused by another group's malevolent magic (Kennedy 1971: 43). Consequently, two of the men of the offending group were killed and their wives taken as booty (*ibid.*). In a place such as prehistoric Britain where the division between the sacred and secular worlds was somewhat blurred, it is not unlikely that some armed conflicts arose because of similar beliefs to those held by the Arunta.

Undoubtedly, another common cause of war in non-sate societies was the lack of a dominant centralised power or state, which meant that for whatever reasons, those who wished to do so, could wage war on other groups without fear of prevention or punishment (Helbling 2006: 121). Thus this somewhat anarchic state of affairs led to what has been termed (Gat 2006: 97) the 'security dilemma' with different groups forced into warfare with each other because groups other than one's own had to be regarded as potential enemies; their very existence posed a threat because they could decide to attack at any time.

The desire of warriors to display their courage and bravery – and thus enhance their reputation and social standing – also acted as a significant factor in non-state warfare. As has been pointed out (Maschner & Reedy Maschner 1998: 22) young males take part in the undeniably risky activity of warfare because they realise that by doing so, they may be placed in a better social or economic position. The Cheyenne of the American Plains provide a much cited, but nevertheless illuminating example of this social elevation through warfare. The highest war honour among them was awarded for *counting coup*. This practice involved a ranked system of bravery, whereby the highest honour was achieved by simply touching an enemy with a *coup* stick; the level of honour then descended in order with the use of the war club, lance bow and arrow and finally the gun (Kennedy 1971: 46). Stealing a single horse from an enemy camp was also seen as an act of great courage (*ibid.*).

Of course, as Richard Osgood (1998: 5) has rightly stated 'we must be careful not to assume a one to one relationship' exists between pre-industrial societies documented by anthropologists and the prehistoric ones recorded by archaeologists. Thus it must not be assumed that the accounts garnered by anthropologists from around the world provide us with a mirror that reflects back the reality of life in prehistoric Britain. Nonetheless, bearing this caveat in mind, this author would agree with Rick Schulting & Michael Wysocki (2005:

107) who have argued that as both ethnographic and historic accounts reveal the prevalence and importance of warfare and violence, 'Then there is no reason to think it was otherwise in prehistory'.

Living in the dangerous world of the twenty-first century, it is understandable that some people hark back to the prehistoric period in Britain as a time when people lived in pristine rural landscapes, their lives untouched by the detritus and dangers of modern life. However, the evidence considered within the pages of this book suggests that such views are misguided and that warfare often raised its ugly head to blight many of Britain's prehistoric communities. Some may argue that the armed conflicts that undoubtedly took place in prehistoric Britain cannot be classed as 'true' warfare, as they were of little consequence and had a minor effect on life in general. It is highly likely however, that if the victims of these conflicts could speak to us now, they would disagree.

BIBLIOGRAPHY

Adkins, L. & Needham, S. 1985. New Research on a Late Bronze Age Enclosure at Queen Mary's Hospital, Carshalton. *Surrey Archaeological Collections* 76, 11-51

Albrethsen, S.E. & Brinch Petersen, E. 1976. Excavation of a Mesolithic Cemetery at Vedbæk, Denmark. *Acta Archaeologica* 47, 1-161

Alcock, L. 1960. Castell Odo: an embanked settlement on Mynydd Ystum, near Aberdaron, Caernarvonshire. *Archaeologia Cambrensis* 109, 78-135

Aldhouse Green, M. 2002. *Dying for the Gods: Human Sacrifice in Iron Age and Roman Europe*. Stroud, Tempus

Aldhouse Green, M. 2004. Chaining and Shaming: Images of Defeat, From Llyn Cerrig Bach to Sarmitzegetusa. *Oxford Journal of Archaeology* 23, 319-340

Anthony, D.W. 1990. Migration in Archaeology: The Baby and the Bathwater. *American Anthropologist* 92, 895-914

ApSimon, A. 1973. Ballnagilly. *Current Archaeology* 3, 11-13

Armit, I. 2003. *Towers in the North: The Brochs of Scotland*. Stroud, Tempus

Armit, I., Murphy, E., Nelis, E. & Simpson, D. (eds) 2003. *Neolithic Settlement in Ireland and Western Britain*. Oxford, Oxbow

Armit, I., Knüsel, C., Robb, J. & Schulting, R. (eds) 2006. Warfare and Violence in Prehistoric Europe: An Introduction. *Journal of Conflict Archaeology* 2, 1-11.

Armit, I. 2007. Hillforts at War: From Maiden Castle to Tainwaha Pā. *Proceedings of the Prehistoric Society* 73, 25-39

Ashbee, P. 1960. *The Bronze Age Round Barrow in Britain*. London, Phoenix House

Ashbee, P. 1984. *The Earthen Long Barrow in Britain*. Norwich, J.M. Dent

Atkinson, J.C. & Evans, J.G. 1978. Recent excavations at Stonehenge. *Antiquity*, 235-236

Avery, M. 1986. 'Stoning and Fire' at hillfort entrances of southern Britain. *World Archaeology* 18, 216-230

Avery, M. 1993. *Hillfort Defences of Southern Britain*. Oxford, British Archaeological Reports 231

Bachechi, L., Fabbri, P.F. & Mallegni, F. 1997. An Arrow-Caused Lesion in a Late Upper Palaeolithic Human Pelvis. *Current Anthropology* 38, 135-140

Bahn, P. (ed.) 1998. *Tombs, Graves and Mummies*. London, Phoenix Illustrated.

Bahn, P. 1999. *Journey Through the Ice Age*. London, Seven Dials

Bahn, P. (ed.) 2002. *Written in Bones: How Human Remains Unlock the Secrets of the Dead*. Devon, David & Charles

Bamforth, D.B. 2004. Indigenous People, Indigenous Violence: Precontact Warfare on the North American Great Plains. *Man* 29, 95-115

Barber, J.W. & Crone, B.A. Crannogs; a diminishing resource? A survey of the crannogs of southwest Scotland and excavations at Buiston Crannog. *Antiquity* 67, 520-533

Barber, M. 2003. *Bronze and the Bronze Age: Metalwork and Society in Britain c.2500 – 800 BC*. Stroud, Tempus

Barclay, A. & Halpin, C. 1999. *Excavations at Barrow Hills, Radley, Oxfordshire. Volume 1: The Neolithic and Bronze Age Monument Complex*. Oxford, Oxford Archaeological Unit

Barclay, G.J., Brophy, K. & MacGregor, G. 2002. A Neolithic building at Claish Farm, near Callander, Stirling Council, Scotland, UK. *Antiquity* 76, 23-24

Barham, L., Priestley, P. & Targett, A. 1999. *In Search of Cheddar Man*. Stroud, Tempus

Barker, G. 1985. *Prehistoric Farming in Europe: its origins and development*. Cambridge University Press, Cambridge

Barton, N. 2005. *Ice Age Britain*. London, BT Batsford

Bartlett, J.E. & Hawkes, C.F.C. 1965. A barbed bronze spearhead from North Ferriby, Yorkshire, England. *Proceedings of the Prehistoric Society* XXXI, 370-373

Bar-Yosef, O. 2004. Guest Editorial: East to West – Agricultural Origins and Dispersal into Europe. *Current Anthropology* 45, 1-3

Beamish, M. & Ripper, S. 2000. Burnt Mounds in the East Midlands. *Antiquity* 74, 37-38

Bellwood, P. 2001. Early Agriculturalist Population Diasporas? Farming, Languages, and Genes. *Annual Review of Anthropology* 30, 181-207

Benson, D. & Whittle, A. 2007. *Building Memories: The Neolithic Cotswold Long Barrow at Ascott-Under-Wychwood, Oxfordshire*. Oxford, Oxbow

Benton, S. 1931. The Excavations of the Sculptor's Cave, Covesea, Morayshire. *Proceedings of the Society of Antiquaries of Scotland* 65, 177-216

Berresford Ellis, P. 2003. *A Brief History of the Celts*. London, Constable & Robinson

Billman, B.R., Lambert, P.M., & Banks, L.L. 2000. Cannibalism, Warfare and Drought in the Mesa Verde region during the Twelfth Century A.D. *American Antiquity* 65, 147-178

Bird, H. 1865. An Account of the Human Bones Found in the Round and Long Tumuli, Situated on the Cotswold Hills near Cheltenham. *Journal of the Anthropological Society of London* 3, 65-74

Blick, G. 1988. Genocidal Warfare in Tribal Societies as a Result of European-Induced Culture Conflict. *Man* (New Series) 23, 654-670

Bogucki, P. 1998. The Talheim Neolithic Mass Burial. In P. Bahn (ed.), *Tombs, Graves and Mummies*, 48-50. London, Phoenix Illustrated

Bradley, R. 1972. Prehistorians and Pastoralists in Neolithic and Bronze Age England. *World Archaeology* 4, 192-204

Bradley, R. 1998. *The Passage of Arms: An Archaeological analysis of prehistoric hoard and votive deposits*. Oxford, Oxbow

Bradley, R. & Ellison, A. 1975. *Rams Hill*. Oxford, British Archaeological Reports

Bradley, R. & Gordon, K. 1988. Human skulls from the Thames, their dating and significance. *Antiquity* 62, 503-509

Breese, C.E. 1932. Castell Odo. *Archaeologia Cambrensis* 1932, 372-386

Bricker, H.M. 1976. Upper Palaeolithic Archaeology. *Annual Review of Anthropology* 5, 133-148

British Archaeology Magazine 2008 (99). *Was missing body a Dutchman in Scotland?* 6

Britton, D. 1963. Traditions of Metal-working in the Later Neolithic and Early Bronze Age of Britain: Part 1. *Proceedings of the Prehistoric Society* XXIX, 258-326

Brothwell, D.R. 1961. An Upper Palaeolithic Skull from Whaley Rock Shelter No. 2, Derbyshire. *Man* 61, 113-116

Brown, P., 1978. *Highland Peoples of New Guinea*. Cambridge, Cambridge University Press.

Brown, A.G. & Edmonds, A.R. (eds) 1987. *Lithic Analysis and Later British Prehistory: Some problems and approaches*. Oxford, British Archaeological Reports 162

Bryant Finney, J. 1996. *Middle Iron Age Warfare of the Hillfort Dominated Zone c. 400 BC to c. 150 BC*. Oxford, British Archaeological Reports 423

Bullock, P. 2002. Kennewick Man. In P. Bahn (ed.), *Written in Bones: How Human Remains Unlock the Secrets of the Dead*, 77-82. Devon, David & Charles

Burl, A. 1981. *Rites of the Gods*. London, J.M. Dent

Burl, A. 1989. *The Stonehenge People*. London, J.M. Dent

Burgess, C., Topping, P., Mordant, C. & Maddison, M. (eds) 1988. *Enclosures and Defences in the Neolithic of western Europe* (i&ii). Oxford, British Archaeological Reports, International Series 403

Burgess, C., Topping, P. & Lynch, F. (eds) 2007. *Beyond Stonehenge: Essays on the Bronze Age in Honour of Colin Burgess*. Oxford, Oxbow

Brück, J. 1995. A place for the dead: the role of human remains in Late Bronze Age Britain. *Proceedings of the Prehistoric Society* 61, 245-279

Carman, J. & Harding, A. (eds) 1999. *Ancient Warfare: Archaeological Perspectives*. Stroud, Sutton

Case, H. 1969. Neolithic Explanations. *Antiquity* XLIII, 176-186

Campbell, C. 1986: Images of War: A problem in San Rock Art Research. *World Archaeology* 18, 255-268

Chadwick, N. 1971. *The Celts*. London, Penguin

Champion, T.C. & Collis, J.R. (eds) 1996. *The Iron Age in Britain and Ireland: Recent Trends*. Sheffield, J.R. Collis

Chapman, J. 1999. The Origins of warfare in the prehistory of Central and Eastern Europe. In J. Carman & A. Harding (eds), 101-143. Stroud, Sutton

Chatters, J.C. 2000. The Recovery and First Analysis of an Early Holocene Human Skeleton from Kennewick, Washington. *American Antiquity* 65, 291-316

Childe, V.G. 1958. *The Prehistory of European Society*. Middlesex, Penguin

Christensen, J. 2004. Warfare in the European Neolithic. *Acta Archaeologica* 75, 129-156

Clark, J.G.D. 1938. A Neolithic House at Haldon, Devon. *Proceedings of the Prehistoric Society* IV, 222-223

Clark, J.G.D. 1954. *Excavations at Starr Carr: An Early Mesolithic site near Scarborough, Yorkshire*. Cambridge, Cambridge University Press

Clark, J.G.D. 1955. Notes on the Obanian with Special Reference to Antler- and Bone-Work. *Proceedings of the Royal Antiquaries of Scotland* 89, 91-106

Clark, J.G.D. 1963. Neolithic Bows from Somerset, England and the Prehistory of Archery in North-Western Europe. *Proceedings of the Prehistoric Society* XXIX, 50-98

Clark, J.G.D. 1966. The Invasion Hypothesis in British Archaeology. *Antiquity* XL, 172-189

Close-Brooks, R. 1995. Excavation of a cairn at Cnip, Uig, Isle of Lewis. *Proceedings of the Society of Antiquaries of Scotland* 125, 253-277

Coles, J. 1962. European Bronze Age Shields. *Proceedings of the Prehistoric Society* XXVIII, 191-209

Coles, J. & Taylor, J. 1971. The Wessex Culture: a minimal view. *Antiquity* XLV, 1-13

Coles, J.M., Hibbert, F.A. & Orme, B.J. 1973. Prehistoric Roads and Tracks in Somerset: 3. The Sweet Track. *Proceedings of the Prehistoric Society* 39. 256-294

Coles, J.M., Heal, V.E. & Orme, B.J. 1978. The use and character of wood in prehistoric Britain and Ireland. *Proceedings of the Prehistoric Society* 44, 1-47

Coles, J.M. & Orme, B.J. 1980. *Prehistory of the Somerset Levels.* Stephen Austin & Sons, Hertford

Corcoran, J.X.W.P. 1964-1966. The excavation of three chambered cairns at Loch Calder, Caithness. *Proceedings of the Society of Antiquaries of Scotland* XCVIII, 1-75

Conklin, B.A. 1995. "Thus Are Our Bodies, Thus Was Our Custom": Mortuary Cannibalism in an Amazonian Society. *American Ethnologist* 22, 75-101

Cowie, T. & Macleod, M. 2002. Lewis Man: A Face from the Past. In P. Bahn (ed.), *Written in Bones*, 32-35. Devon, David & Charles

Cunliffe, B. 1992. Pits, Preconceptions and Propitiation in the British Iron Age. *Oxford Journal of Archaeology* 11, 69-83

Cunliffe, B. (ed.) 1997. *Prehistoric Europe: an Illustrated History.* Oxford, Oxford University Press

Cunliffe, B. 2003. *Danebury Hillfort.* Stroud, Tempus

Cunliffe, B. 2004a. Wessex Cowboys? *Oxford Journal of Archaeology* 23, 61-81

Cunliffe, B. 2004b. *Iron Age Britain.* London, BT Batsford

Cunliffe, B. 2005. *Iron Age Communities in Britain.* Oxon, Routledge

Currant, A.P., Jacobi, R.M. & Stringer, C.B. 1989. Excavations at Gough's Cave, Somerset 1986-87. *Antiquity* 63, 131-136

Current Archaeology Magazine 2007 (209). *The New Radiocarbon Dating Revolution.* 9-20

Curwen, E.C. 1954. *The Archaeology of Sussex.* London, Methuen & Co.

Darole, R. & Divale, W.T. 1976. Natural Selection in Cultural Evolution: Warfare versus

Peaceful Diffusion. *American Ethnologist* 3, 97-131

Darvill, T. 1987. *Prehistoric Britain.* London, B.T. Batsford

Darvill, T. 2003. Billown and the Neolithic of the Isle of Man. In I. Armit, E. Murphy, E. Nelis & D. Simpson (eds), *Neolithic Settlement in Ireland and Western Britain*, 112-120. Oxford, Oxbow

Davies, D. 1857. Celtic Sepulture on the Mountains of Carno, Montgomeryshire. *Archaeologia Cambrensis* III, 301-305

Defleur, A., White, T., Valensi, P., Slimak, L. & Crégut-Bonnoure, É. 1999. Neanderthal Cannibalism at Moula-Guercy, Ardèche, France. *Science* 286, 128-131

Dennell, R.W. 1984. The Expansion of Exogenous-Based Economies Across Europe: The Balkans and Central Europe. In S. De Atley & F.J. Findlow (eds), *Exploring the Limits: Frontiers and Boundaries in Prehistory*, 93-117. Oxford, British Archaeological Reports, International Series 223

Dent. J.S. 1983. Weapons, Wounds and War in the Iron Age. *Archaeological Journal* 140, 120-128

Dent, J. 1985. Three cart burials from Wetwang, Yorkshire. *Antiquity* LIX, 85-92

Dickinson, O. 1994. *The Aegean Bronze Age.* Cambridge, Cambridge University Press

Dixon, P. & Borne, P. 1977. *Crickley Hill and Gloucestershire in Prehistory.* Gloucester, Crickley Hill Trust & Gloucestershire County Council

Dixon, P. 1988. The Neolithic settlements on Crickley Hill. In C. Burgess, P. Topping, C. Mordant. & M. Maddison (eds), *Enclosures and Defences in the Neolithic of Western Europe* (i), 75-89. Oxford, British Archaeological Reports, International Series 403

Donaldson, P. 1977. The Excavation of a Multiple Round Barrow at Barnack, Cambridgeshire, 1974-1976. *Antiquaries Journal* LVII, 197-232

Dunwell, A.J., Neighbour, T. & Cowie, T.G. 1995. A cist burial adjacent to the Bronze Age cairn at Cnip, Uig, Isle of Lewis. *Proceedings of the Society of Antiquaries of Scotland* 125, 279-288

Dyer, J. 1992. *Hillforts of England and Wales.* Princes Risborough, Shire

Earnshaw, J.R. 1973. The Site of a Medieval Post Mill and Prehistoric site at Bridlington. *The Yorkshire Archaeological Journal* 45, 19-41

Edmonds, M. & Thomas, J. 1987. The Archers: An Everyday Story of Country Folk. In A.G. Brown & M. Edmonds (eds), *Lithic Analysis and Later British Prehistory: Some problems and approaches*, 187-199. Oxford, British Archaeological Reports 162

Edmonds, M. 1995. *Stone tools and Society: Working Stone in Neolithic and Bronze Age Britain.* London, B.T. Batsford

Edwards, K.J. & Ralston, I.B.M. 2003. *Scotland after the Ice Age: Environment, Archaeology and History, 8000 BC – AD 1000.* Edinburgh, Edinburgh University Press.

Ehenreich, B.1997. *Blood Rites: Origins and History of the Passions of War.* London, Virago

Ellis, P. 1989. Norton Fitzwarren hillfort: a report on the excavations by Nancy and Phillip Langmaid between 1968 and 1971, *Somerset Archaeology and Natural History* 133, 1-74

Fairweather, A.D. 1993. The Neolithic timber hall at Balbridie, Grampian Region, Scotland: the building, the date, and the plant macrofossils. *Antiquity* 76, 313-323

Ferguson, R.B. 1984. Introduction: studying war. In R.B. Ferguson (ed.), *Warfare, Culture, and Environment*, 1-81. Academic Press, Orlando

Ferguson, R.B. 1997. Review of War before Civilization, in *American Anthropologist* 99, 424-425

Fernández-Jalvo, Y., Díez, J.C., Cáceres, I. & Rossel, J. 1999. Human cannibalism in the early Pleistocene of Europe. *Journal of Human Evolution* 37, 597-622

Fewster, K.J. & Zvelebil, M. (eds) 2001. *Ethnoarchaeology and Hunter-Gatherers: Pictures at an Exhibition.* Oxford, British Archaeological Reports, International Series 955

Fitzpatrick, A. 2003. The Amesbury Archer. *Current Archaeology* 184, 146-152

Fokkens, H., Achterkamp, Y. & Kuijpers, M. 2008. Bracers or bracelets? About the functionality and meaning of Bell-beaker wrist-guards. *Proceedings of the Prehistoric Society* 74, 109-140

Fleming, A. 1971. Territorial Patterns in Bronze Age Wessex. *Proceedings of the Prehistoric Society* XXXVII, 138-167

Fleming, A. 1978. The prehistoric landscape of Dartmoor Part I: South Dartmoor. *Proceedings of the Prehistoric Society* 44, 97-123

Fyllingen, H. 2006. Society and the Structure of Violence: A Story Told by Middle Bronze Age Human Remains from Central Norway. In T. Otto, H. Thrane & H. Vankilde (eds), *Warfare and Society: Archaeological and Social Anthropological Perspectives*, 341-385. Denmark, Aarhus University Press

Galer, D. & Knüsel, C. 2007. The Human Remains. In D. Benson and A. Whittle (eds.), *Building Memories: The Neolithic Cotswold Long at Ascott-Under-Wychwood, Oxford*, 189-220. Oxford, Oxbow

Garton, D. 1987. Buxton. *Current Archaeology* 103, 250-253

Gat, A. 1999. The Pattern of Fighting in Simple, Small-Scale, Prestate Societies. *Journal of Anthropological Research* 55, 563-583

Gat, A. 2006. *War in Human Civilization*. Oxford, Oxford University Press

Geus, F. 2002. *The Middle Nile Valley from Later Prehistory to the end of the New Kingdom*. Paper given at the International Society for Nubian Studies 10th International Conference

Gibson, A. 1992. The timber circle at Sarn-y-Bryn-Caled, Welshpool, Powys: ritual and sacrifice in Bronze Age mid-Wales. *Antiquity* 66, 84-92

Gibson, A. 1994. Excavations at the Sarn-y-bryn-caled Cursus Complex, Welshpool, Powys, and the timber circles of Great Britain and Ireland. *Proceedings of the Prehistoric Society* 60, 143-225

Gibson, A. (ed) 2002. *Behind Wooden Walls: Neolithic Palisaded Enclosures in Europe*. Oxford, British Archaeological Reports, International Series 1013

Gilchrist, R. 2003. Introduction: towards a social archaeology of warfare. *World Archaeology* 35, 1-6

Glenn, T.A. 1914. Exploration of a Neolithic Station near Gwaenysgor, Flintshire. *Archaeologia Cambrensis* XIV, 247-270

Golitko, M. & Keeley, L.H. 2007. Beating ploughshares back into swords: warfare in the *Linearbandkeramik*. *Antiquity* 81, 333-342

Green, H.S. 1980. *The Flint Arrowheads of the British Isles*. Oxford, British Archaeological Reports 75

Grimes, W.F. 1938 A Barrow on Breach Farm, Llanbleddian, Glamorgan. *Proceedings of the Prehistoric Society* IV (1), 107-122

Grinsell, L. 1941. The Bronze Age Round Barrows of Wessex. *Proceedings of the Prehistoric Society* VII, 73-114

Guttmann, E.B.A. & Last, J. 2000. A Late Bronze Age Landscape at South Hornchurch, Essex. *Proceedings of the Prehistoric Society* 66, 319-361

Hallpike, C. 1977. *Bloodshed and Vengeance in the Papuan Mountains*. Oxford, Oxford University Press

Hamilton, S. & Manley, J. 2001. Hillforts, Monumentality and Place: A Chronological and Topographic Review of First Millennium BC Hillforts of South-East England. *European Journal of Archaeology* 4, 7-42

Harbison, P. 1971. Wooden and Stone Chevaux-de-frise in Central and Western Europe. *Proceedings of the Prehistoric Society* XXXVII, 195-226

Harding, A. 2000. *European Societies in the Bronze Age*. Cambridge, Cambridge University Press

Harrison, R.J. 1980. *The Beaker Folk*. London, Thames & Hudson

Hayano, D.M. 1974. Marriage, Alliance and Warfare: A View from the New Guinea Highlands. *American Ethnologist* 1, 281-293

Heath, J. 2006. *Ancient Echoes: the early history of a Welsh Peninsula*. Llanrwst, Gwasg Carreg Gwalch

Hedges, J. & Buckley, D. 1978. Excavations at a Neolithic causewayed enclosure, Orsett, Essex, 1975. *Proceedings of the Prehistoric Society* 44, 219-309

Hedges, J.W. & Bell, B. 1980. That Tower of Scottish prehistory – the broch. *Antiquity* LIV, 87-94

Hedges, R.E.M., Housley, R.A., Bronk, C.R. & Vanklinken, G.J. 1991. Radiocarbon Dates from the Oxford AMS System: Archaeometry Datelist 12. *Archaeometry* 33, 121-134

Helbling, J. 2006. War and Peace in Societies without Central Power: Theories and Perspectives. In T. Otto, H. Thrane & H. Vankilde (eds), *Warfare and Society:* Sociological and Social Anthropological Perspectives, 113-141. Denmark, Aarhus University Press

Henderson, J.C. 1998. Islets Through Time: The Definition, Dating and Distribution of Scottish Crannogs. *Oxford Journal of Archaeology*, 227-244

Henderson, J.C. (ed.) 2000. *The Prehistory and Early History of Atlantic Europe.* Oxford, British Archaeological Reports, International Series, S861

Hendrickx, S. & Vermeersch, P. 2000. Prehistory: From the Palaeolithic to the Badarian Culture (c.700,000-4000 BC). In I. Shaw (ed.), *The Oxford History of Ancient Egypt*, 17-44. Oxford, Oxford University Press

Hill, J.D. 1996. Hill-forts and the Iron Age of Wessex. In T.C. Champion & J.R. Collis (eds), *The Iron Age in Britain and Ireland: Recent Trends.* Sheffield, J.R. Collis

Hoffman, M.A. 1984. *Egypt Before the Pharaohs.* London, ARK

Houlder, C.H. 1963. A Neolithic Settlement on Hazard Hill, Totnes. *Proceedings of the Devon Archaeological Exploration Society* 21, 2-31

Ives, S. 2003. Was Ancient Alpine "Iceman" Killed in Battle? *National Geographic News*

James, S. & Rigby, V. 1997. *Britain and the Celtic Iron Age.* London, British Museum Press

Jiménez, G.A & Sánchez Romero. 2006. The origins of warfare: later prehistory in southeastern Iberia. In M. Parker Pearson & I.J.N. Thorpe (eds), *Warfare, Violence and Slavery in Prehistory.* Oxford, British Archaeological Reports, International Series 1374.

Johnston, D.E. 1980. The Excavation of a Bell-Barrow at Sutton Veny, Wilts. *Wiltshire Archaeological Magazine* 72-73, 29-51

Jope, E.M. 1961. Daggers of the Early Iron Age in Britain. *Proceedings of the Prehistoric Society*, 307-344

Keeley, L.H., & Cahen, D. 1989. Early Neolithic Forts and Villages in Belgium: A Preliminary Report. *Journal of Field Archaeology* 16, 157-176

Keeley, L.H. 1996. *War Before Civilization: The Myth of the Peaceful Savage.* Oxford, Oxford University Press

Keeley, L.H., Fontana, M., & Quick, R. 2007. Baffles and Bastions: The Universal Features of Fortifications. *Journal of Archaeological Research* 15, 55-95

Kennedy, J.G. 1971. Ritual and Intergroup Murder: Comments on War, Primitive and Modern. In M.N. Walsh (ed.), *War and the human race*, 41-61. London, Elsevier

Keyser, J.D. 1979. The Plains Indian War Complex and the Rock Art of Writing-on-Stone, Alberta, Canada. *Journal of Field Archaeology* 6, 41-48

Kim, J. 2003. Land-Use Conflict and the Rate of the Transition to Agricultural Economy: A Comparative Study of Southern Scandinavia and Central-Western Korea. *Journal of Archaeological Method and Theory* 10, 277-320

Kinnes, I.A. & Longworth, I.H. 1985. *Catalogue of the Excavated Prehistoric and Romano-British Material in the Greenwell Collection.* London, British Museum Publications

Knight, R.W., Browne, C. & Grinsell, L.V. 1972. Prehistoric Skeletons from Tormarton. *Transactions of the Bristol and Gloucestershire Archaeological Society* XCI, 14-18

Kristiansen, K. 2002. The Tale of the Sword – Swords and Swordfighters in Bronze Age Europe. *Oxford Journal of Archaeology* 21, 319-332

Lillie, M.C. 2001. Mesolithic cultures of Ukraine: Observations on cultural developments in light of new radiocarbon determinations from the Dnieper Rapids cemeteries. In J.C. Henderson (ed.), *Ethnoarchaeology and Hunter-Gatherers: Pictures at an Exhibition*, 53-65. Oxford, British Archaeological Reports, International Series 955

Lillie, M.C. 2004. Fighting for your Life? Violence at the Late-glacial to Holocene transition in Ukraine. In M. Roksandic (ed.), *Violent Interactions in the Mesolithic: Evidence and Meaning*, 89-97. Oxford, British Archaeological Reports

Lindenbaum, S. 2004. Thinking about Cannibalism. *Annual Review of Anthropology* 33, 475-498

Logue, P. 2003. Excavations at Thornhill, Co. Londonderry. In I. Armit, E. Murphy, E. Nelis & D. Simpson (eds), *Neolithic Settlement in Ireland and Western Britain*, 149-156. Oxford, Oxbow

Loveday, R. 2002. Duggleby Howe Revisited. *Oxford Journal of Archaeology* 21, 135-146

Lukaschek, K. 2000/2001. *The History of Cannibalism*. Unpublished MPhil thesis, Cambridge University

Lynch, F. 1970. *Prehistoric Anglesey*. Llangefni, Anglesey Antiquarian Society

Lynch, F. & Burgess, C. (eds) 1972. *Prehistoric Man in Wales and the West*. Bath, Adams & Dart

Lynch, F., Aldhouse-Green, S. & Davies, J.L. 2000. *Prehistoric Wales*. Stroud, Sutton

McGrail, S. 1979. Prehistoric Boats, Timber, and Woodworking Technology. *Proceedings of the Prehistoric Society* 45, 159-163

Mackie, E.W. 1965. Brochs and the Hebridean Iron Age. *Antiquity* XXXIX, 266-278

Mackie, E.W. 2008. The Broch Cultures of Atlantic Scotland: Origins, High Noon and Decline. Part 1: Early Iron Age Beginnings *c*.700-200 BC. *Oxford Journal of Archaeology* 27, 261-279

Manby, T.G. 2007. Continuity of monumental traditions into the Bronze Age? Henges to ring-forts, and shrines. In C. Burgess, P. Topping & F.Lynch (eds), *Beyond Stonehenge: Essays on the Bronze Age in Honour of Colin Burgess*, 403-425. Oxford, Oxbow

Maschner, D.G. & Reedy-Maschner, K.L. 1998. Raid, Retreat, Defend (Repeat): The Archaeology and Ethnohistory of Warfare on the North Pacific Rim. *Journal of Anthropological Archaeology* 17, 19-51

Megaw, B.R.S. & Hardy, E.M. 1938. British Decorated Axes and their Diffusion during the Earlier Part of the Bronze Age. *Proceedings of the Prehistoric Society* IV (2), 272-308

Megaw, J.V.S & Simpson, D.D.A. (eds) 1979. *Introduction to British Prehistory*. Leicester & London, Leicester University Press

Meggit, M. 1977. *Blood is their Argument: warfare among the Mae Enga tribesmen of the New Guinea highlands*. Mayfield, Palo Alto, California

Mercer, R. 1981. Excavations at Carn Brea, Illogan, Cornwall – a Neolithic Fortified Complex of the Third Millennium bc. *Cornish Archaeology* 20, 1-205

Mercer, R. 1986. The Neolithic in Cornwall. *Cornish Archaeology* 25, 35-81

Mercer, R. 1988. Hambledon Hill, Dorset, England. In C. Burgess, P. Topping, C. Mordant & C. Maddison (eds), Enclosures and Defences in the Neolithic of Western Europe (i), 89-107. Oxford, British Archaeological Reports, International Series 403,

Mercer, R. 1990. *Causewayed Enclosures*. Princes Risborough, Shire

Mercer, R. 1999. The Origins of Warfare in the British Isles. In J. Carman & A. Harding (eds), *Ancient Warfare: Archaeological perspectives*, 143-157. Stroud, Sutton

Mercer, R. 2006. By Other Means? The Development of Warfare in the British Isles 3000-500 BC. In I. Armit, C. Knusel, J. Robb & R. Schulting (eds), *Journal of Conflict Archaeology* 2, 119-151
Midant-Reynes, B. 2000.The Naqada Period (c.4000-3200 BC). In I. Shaw (ed.), *The Oxford History of Ancient Egypt*, 44-61. Oxford, Oxford University Press
Miller, M., McEwen, E. & Bergman, C. 1986. Experimental approaches to ancient Near Eastern archery. *World Archaeology* 18, 178-195
Mithen, S.J. 1997. The Mesolithic Age. In B. Cunliffe (ed.) *Prehistoric Europe: An Illustrated History*, 79-136. Oxford, Oxford University Press
Mohen, J-P. 1990. *The World of the Megaliths*. Oxford, Facts On File
Mohen, J-P. & Eluère, C. 2000. *The Bronze Age in Europe*: Gods, Heroes and Treasures. London, Thames & Hudson
Molloy, B. 2008. Martial arts and materiality: a combat archaeology perspective on Aegean swords of the fifteenth and fourteenth centuries BC. *World Archaeology* 40, 116-134
Nash, G. 2005. Assessing rank and warfare-strategy in prehistoric hunter-gatherer society: a study of representational warrior figures in rock-art from the Spanish Levant, southeastern Spain. In M. Parker Pearson & I.J.N. Thorpe (eds), *Warfare, Violence and Slavery in Prehistory*, 75-87. Oxford, British Archaeological Reports, International Series 1374
Needham, S. 1979. Two Recent British Shield Finds and their Continental Parallels. *Proceedings of the Prehistoric Society* 45, 111-135
Needham, S. 1990. Middle Bronze Age Ceremonial Weapons: New Finds from Oxborough, Norfolk and Essex/Kent. *The Antiquaries Journal* LXX, 239-253
Needham, S. & Ambers, J. 1994. Redating Rams Hill and reconsidering Bronze Age enclosure. *Proceedings of the Prehistoric Society* 60, 225-245
Needham, S., Pitts, M., Heyd, V., Parker Pearson, M., Jay, M., Montgomery, J. & Sheridan, S. 2008a. In the Copper Age. *British Archaeology* 101, 19-27
Needham, S., Lawson, A. & Woodward, A. 2008b. Rethinking Bush Barrow. *British Archaeology* 104, 12-18
Newcomb, Jr, W.W. 1950. A Re-Examination of the Causes of Plains Warfare. *American Anthropologist* 3, 317-330
Newell, R.R., Constandse-Westerman, T.S & Meiklejohn, C. 1979. The skeletal remains of Mesolithic man in western Europe: an evaluative catalogue. *Journal of Human Evolution* 8, 1-228
Noble, G. 2003. Islands and the Neolithic Farming Revolution. *British Archaeology* 71, 21-22
Orschiedt, J. 2005. The Head Burials from Ofnet Cave: an example of warlike conflict in the Mesolithic. In M. Parker Pearson (eds), Warfare, Violence and Slavery in Prehistory, 67-73. Oxford, British Archaeological Reports, International Series 1374
Osgood, R. 1998. *Warfare in the Late Bronze Age of North Europe*. Oxford, British Archaeological Reports (International series) 694
Osgood, R. 1999. Britain in the age of warrior heroes. *British Archaeology* 46, 8-10
Osgood, R., Monks, S. & Toms, J. 2000. Bronze Age Warfare. Stroud, Sutton
Osgood, R. 2005. The dead of Tormarton – Middle Bronze Age combat victims? In M. Parker Pearson & I.J.N. Thorpe (eds), *Warfare, Violence and Slavery in Prehistory*, 139-145. Oxford, British Archaeological Reports, International Series 1374
Osgood, R. 2006. The dead of Tormarton: Middle Bronze Age combat victims? In T. Otto, H. Thrane & H. Vankilde (eds), *Warfare and Society: Archaeological and Social Anthropological Perspectives*, 331-341. Denmark, Aarhus University Press

Otterbein, K.F. 1968. Internal War: A Cross-Cultural Study. *American Anthropologist* 70, 277-289

Otterbein, K.F. 1999. A History of Research on Warfare in Anthropology. *American Anthropologist* 101, 794-805

Otterbein, K.F. 2000. The Killing of Captured Enemies: A Cross-Cultural Study. *Current Anthropology* 41, 439-443

Otto, T. 2006. Warfare and Exchange in a Melanesian Society before Colonial Pacification: The Case of Manus, Papua New Guinea. In T. Otto, H. Thrane & H. Vankilde (eds), *Warfare and Society: Archaeological and Social Anthropological Perspectives*, 187-201. Denmark, Aarhus University Press

Oxford Archaeological Unit. 2000. White Horse Stone: A Neolithic Longhouse. *Current Archaeology* 168, 450-453

Parker Pearson, M. 1993. *Bronze Age Britain*. London, B.T. Batsford/English Heritage

Parker Pearson, M. & Thorpe, I.J.N. (eds) 2005. *Warfare, Violence and Slavery in Prehistory*. Oxford, British Archaeological Reports, International Series 1374

Parkinson, W.A. & Duffy, P.R. 2007. Fortifications and Enclosures in European Prehistory: A Cross-Cultural Perspective. *Journal of Archaeological Research* 15. 97-141

Parry, F., Davies, J.A., St. George Gray, H., Keith, A. & Cooper, N.C. 1928. Excavations at the Caves, Cheddar. *Proceedings of the Somersetshire Archaeological and Natural History Society* LXXVI, 102-122

Parry, R.F., St. George Gray, H., Cooper, N.C. & Wilfrid Jackson, J. 1930. Excavations at Cheddar. *Proceedings of the Somersetshire Archaeological and Natural History Society* LXXVI, 46-63

Passmore, A.D. 1942. Chute, barrow 1. *Wiltshire Archaeological and Natural History Magazine* 50, 100-101

Pettit, P. 1974. *Prehistoric Dartmoor*. Devon, David & Charles

Piggot, S. 1938. The Early Bronze Age in Wessex. *Proceedings of the Prehistoric Society* IV, 52-107

Piggott, S. 1948. The Excavations at Cairnpapple Hill, West Lothian, 1947-48. *Proceedings of the Royal Society of Antiquaries of Scotland* XXXXII, 68-124

Piggot, S. 1950. Swords and Scabbards of the British Early Iron Age. *Proceedings of the Prehistoric Society* XVI, 1-29

Piggott, S. 1954. *The Neolithic Cultures of the British Isles*. Cambridge, Cambridge University Press

Piggott, S. 1959. The Carnyx in Early Iron Age Britain. *The Antiquaries Journal* XXXIX, 19-33

Piggot, S. 1962. *The West Kennet Long Barrow: Excavations 1955-56*. London, HMSO

Piggott, S. 1971. Beaker bows: a suggestion. *Proceedings of the Prehistoric Society* XXXVII (part 2), 80-95

Pitt-Rivers, A.H.L.F. 1898. *Excavations in Cranborne Chase* IV. London, privately printed

Pitts, M. 2001. *Hengeworld*. London, Arrow

Pollard, S. 1966. Neolithic and Dark Age settlements on High Peak, Sidmouth, Devon. *Proceeding of the Devon Archaeological Society* 23, 35-39

Pollard, J. 1997. *Neolithic Britain*. Princes Risborough, Shire

Price, T.D. 1985. *Prehistoric Hunter-Gatherers: The Emergence of Cultural Complexity*. London, Academic Press

Price, T.D. 1985. Affluent Foragers of Mesolithic Southern Scandinavia. In T. D. Price & J.A. Brown (eds), *Prehistoric Hunter-Gatherers: The Emergence of Cultural Complexity*, 341-363. London, Academic Press

Price, T.D. & Gebauer, A.B. (eds) 1995. *Last Hunters-First Farmers: New Perspectives on the Prehistoric Transition to Agriculture*. Mexico, School of American Research Press

Pritchard-Evans, E.E. 1960. Zande Cannibalism. *The Journal of the Royal Anthropological Institute of Great Britain and Ireland* 90, 238-258

Pryor, F. 1976. A Neolithic multiple burial from Fengate, Peterborough. *Antiquity* 50, 232-233

Pryor, F. 2003. *Britain BC: Life in Britain and Ireland before the Romans*. London, Harper Collins

Rebecca Craig, C., Knüsel, C.J., & Carr, G.C. 2005. Fragmentation, mutilation and dismemberment: an interpretation of human remains on Iron Age sites. In M. Parker Pearson & I.J.N. Thorpe (eds), *Warfare, Violence and Slavery in Prehistory*, 165-180. Oxford, British Archaeological Reports, International Series 1374

Ritchie, J.N.G. 1988. *Brochs of Scotland*. Princes Risborough, Shire

Robertson-Mackay, R. 1987. The neolithic causewayed enclosure at Staines, Surrey: excavations 1961-63. *Proceedings of the Prehistoric Society* 53, 23-129

Roksandic, M. (ed.) 2004. *Violent Interactions in the Mesolithic: Evidence and Meaning*. Oxford, British Archaeological Reports, International Series 1237

Roksandic, M. 2004. Contextualising the Evidence of Violent Death in the Mesolithic: Burials associated with Victims of Violence in the Iron Gates Gorge. In M. Roksandic (ed.), *Violent Interactions in the Mesolithic: Evidence and Meaning*, 53-75. Oxford, British Archaeological Reports, International Series 1237

Rowley-Conwy, P. 2004. How the West Was Lost. A Reconsideration of Agricultural Origins in Britain, Ireland and Southern Scandinavia. *Current Anthropology* 45, 83-113

Russel, A.D. 1990. Two Beaker burials from Chilbolton, Hampshire. *Proceedings of the Prehistoric Society* 56, 153-173

Savory, H.N. 1971. A Welsh bronze age hillfort. *Antiquity* XLV, 251-261

Schulting, R.J. 1996. Antlers, bone pins and flint blades: the Mesolithic cemeteries of Téveic and Höedic, Brittany. *Antiquity* 70, 335-350

Schulting, R. & Richards, M. 2002. The wet, the wild and the domesticated: the Mesolithic-Neolithic transition on the west coast of Scotland. *European Journal of Archaeology* 5, 147-189

Schulting, R. & Wysocki, M. 2005. 'In this Chambered Tumulus were found Cleft Skulls ...': an Assessment of the Evidence for Cranial Trauma in the British Neolithic. *Proceedings of the Prehistoric Society* 71, 107-138

Schutz, H. 1983. *The Prehistory of Germanic Europe*. New Haven & London, Yale University Press

Seligman, C.G., & Parsons, F.G. 1914. The Cheddar Man: A Skeleton of Late Palaeolithic Date. *The Journal of the Royal Anthropological Institute of Great Britain and Ireland* 44, 241-263

Selkirk, A. 1971. Ascott-Under-Wychwood. *Current Archaeology* 24, 7-10

Senior, M. 2005. *Hillforts of northern Wales*. Llanrwst, Gwasg Carreg Gwalch

Sharples, N. 1991a. *Maiden Castle: Excavations and field survey 1985-6*. London, English Heritage

Sharples, N. 1991b. *Maiden Castle*. London, Batsford/English Heritage

Shaw, I. (ed.) 2000. *The Oxford History of Ancient Egypt*. Oxford, Oxford University Press

Shepherd, I.A.G. 2007. 'An Awesome Place'. The late Bronze Age use of the Sculptor's Cave, Covesea, Moray. In C. Burgess, P. Topping & F. Lynch (eds), *Beyond Stonehenge: Essays on the Bronze Age in Honour of Colin Burgess*. Oxford, Oxbow

Sheridan, A. 2000 Achnacreebeag and its French Connections: Vive the 'Auld Alliance'. In J.C. Henderson (ed.), *The Prehistory and Early History of Atlantic Europe*, 1-15. Oxford, British Archaeological Reports, International Series 861

Simpson, D.D.A. & Coles, J.M. 1990. Excavations at Grandtully, Perthshire. *Proceedings of the Society of Antiquaries of Scotland* 120, 33-44

Smith, C. 1992. *Late Stone Age Hunters of the British Isles*. London, Routledge

Smith, J. 2006. *Early Bronze Age Stone Wrist-Guards in Britain: archer's bracer or social symbol?* http://www.geocities.com/archchaos1/article/1.htm

Smith, M. & Brickley, M. 2007. Boles Barrow: Witness to Ancient Violence. *British Archaeology* 93, 22-28

Smith, M.J., Brickley, M.B. & Leach, S.L. 2007. Experimental evidence for lithic projectile injuries: improving identification of an under recognised phenomenon. *Journal of Archaeological Science* 34, 540-533

Smith, M.O. 2003. Beyond Palisades: The Nature and Frequency of Late Prehistoric Deliberate Violent Trauma in the Chickamauga Reservoir of East Tennessee. *American Journal of Physical Anthropology* 121, 303-318

Speak, S. & Burgess, C. 1999. Meldon Bridge: a centre of the third millennium BC in Peeblesshire. *Proceedings of the Society of Antiquaries of Scotland* 129, 1-118

Spielmann, K.A. & Eder, J.F. 1994. Hunters and Farmers: Then and Now. *Annual Review of Anthropology* 23, 303-23

Souden, D. 1997. *Stonehenge: Mysteries of the Stones and Landscape*. London, Collins & Brown

Standen, V.S. & Arriaza, B.T. 2000. Trauma in the Preceramic Coastal Populations of Northern Chile: Violence or Occupational Hazards? *American Journal of Physical Anthropology* 112, 239-249

Starling, N.J. 1985. Social Change in the Later Neolithic of Central Europe. *Antiquity* LIX, 30-38

Stead, I. 1988. Chalk Figurines of the Parisi. *The Antiquaries Journal* LXVIII, 9-30

Stead, I.M. 1991. *Iron Age Cemeteries in East Yorkshire: Excavations at Burton Fleming, Rudston, Garton-on-the-Wolds and Kirkburn*. London, English Heritage/British Museum Press

Stead, I.M. 2006. *British Iron Age Swords and Scabbards*. London, British Museum Press

Stevenson, R.B.K. 1960. A wooden sword of the late Bronze Age. *Proceedings of the Society of Antiquaries of Scotland* 91, 191-193

Taçon, P. & Chippendale, C. 1994. Australia's ancient warriors: changing depictions of fighting in the rock-art of Arnhem Land, N.T. *Cambridge Archaeological Journal* 4, 211-248

Taylor, J.J. 1994. The First Golden Age of Europe was in Ireland and Britain (Circa 2400-1400 B.C.). *Ulster Journal of Archaeology* 57, 37-60

Taylor, T. 2001. The Edible Dead. *British Archaeology* 59, 8-14

Thomas, J. 1991. *Understanding the Neolithic*. Oxon, Routledge

Thomas, J. 2004. Current debates on the Mesolithic-Neolithic transition in Britain and Ireland. *Documenta Praehistorica* XXXI, 113-130

Thorpe, I.J. 1996. *The Origins of Agriculture in Europe*. London, Routledge

Thorpe, N. 2000. Origins of War: Mesolithic Conflict in Europe. *British Archaeology* 52, 8-14

Thorpe, I.J.N. 2003. Anthropology, archaeology, and the origin of warfare. *World Archaeology* 35, 145-165

Thorpe, N. 2006. Fighting and Feuding in Neolithic and Bronze Age Britain and Ireland. In T. Otto, H. Thrane and H. Vankilde (eds), *Warfare and Society: Archaeological and Social Anthropological Perspectives*, 147-167. Denmark, Aarhus University Press

Thurnam, J. 1856. On a cromlech-tumulus called Lugbury, near Littleton Drew: And Note on the name of Drew. *Wiltshire Archaeological and Natural History Magazine* 3, 164-177

Tomb, J.J. & Davies, O. 1938. Urns from Ballymacaldrack. *Ulster Journal of Archaeology* 1, 219-221

Tresset, A. 2000. Early Husbandry in Atlantic Areas. Animal Introductions, Diffusions of Techniques and Native Acculturation at the North-Western Fringe of Europe. In J.C. Henderson (ed.), *The Prehistory and Early History of Atlantic Europe*, 17-32. Oxford, British Archaeological Reports, International Series 861

Trump, B.A.V. 1962. The Origin and Development of British Middle Bronze Age Rapiers. *Proceedings of the Prehistoric Society* XXVIII, 80-103

Vankilde, H. 2003. Commemorative tales: archaeological responses to modern myth, politics, and war. *World Archaeology* 35, 126-144

Vilaça, A. 2000. Relations between Funerary Cannibalism and Warfare Cannibalism: The Question of Predation. *Ethnos* 65, 83-106

Vutiropulos, N. 1991. The Sling in the Aegean Bronze Age. *Antiquity* 65, 279-286

Waddell, J. 2000. *The Prehistoric Archaeology of Ireland*. County Wicklow, Wordwell

Wainwright, G.J. 1967. *Coygan Camp: A Prehistoric, Romano-British and Dark Age Settlement in Carmarthenshire*. The Cambrian Archaeological Association

Wainwright, G.J. 1979. *Mount Pleasant, Dorset: Excavations 1970-1971*. London, The Society of Antiquaries

Walker, P.L. 2001. A Bioarchaeological Perspective on the History of Violence. *Annual Review of Anthropology* 30, 573-596

Wheeler, R.E.M. 1943. *Maiden Castle, Dorset*. Oxford, Reports of the Research Committee of the Society of Antiquaries XII

Whitley, J. 2002. Too many ancestors. *Antiquity* 76, 119-126

Whittle, A. 1991a. Wayland's Smithy, Oxfordshire: excavations at the Neolithic tomb in 1962-63 by R.J.C. Atkinson and S. Piggott. *Proceedings of the Prehistoric Society* 57 (2), 61-103

Whittle, A. 1991b. A late Neolithic complex at West Kennet, Wiltshire, England. *Antiquity*, 256-62

Whittle, A. 1993. The Neolithic of the Avebury area: Sequence, Environment, Settlement and Monuments. *Oxford Journal of Archaeology* 12, 29-53

Whittle, A. 1996. *Europe in the Neolithic: The Creation of new worlds*. Cambridge, Cambridge University Press

Wilson, I. 2001. *Past Lives: Unlocking the Secrets of Our Ancestors*. London, Cassel & Co

Woodward, A., Hunter, J., Ixer, R., Roe, F., Potts, P.J., Webb, P.C., Watson, J.S. & Jones, M.C. 2006. Beaker age bracers in England: sources, function and use. *Antiquity* 80, 530-543

Wright, E.V. & Churchill, D.M. 1965. The Boats from North Ferriby, Yorkshire, England, with a review of the origins of the sewn boats of the Bronze Age. *Proceedings of the Prehistoric Society* XXXI, 1-25

Wysocki, M. & Whittle, A. 2000. Diversity, lifestyles and rites: new biological and archaeological evidence from British Earlier Neolithic mortuary assemblages. *Antiquity* 285, 591-601

York, J. 2002. The Life Cycle of Bronze Age metalwork from the Thames. *Oxford Journal of Archaeology* 21, 77-92

INDEX